THE ILLUSTRATED

ON THE SHOULDERS OF GIANTS

THE GREAT WORKS OF PHYSICS AND ASTRONOMY

THE ILLUSTRATED

ON THE SHOULDERS OF GIANTS

THE GREAT WORKS OF PHYSICS AND ASTRONOMY

STEPHEN HAWKING

RUNNING PRESS
PHILADELPHIA • LONDON

A BOOK LABORATORY BOOK

© 2004 by Stephen Hawking
© 2004 original illustrations by The Book Laboratory® Inc.

9 8 7 6 5 4 3 2 1
Digit on the right indicates the number of this printing
Library of Congress Control Number: 20040745

ISBN 0-7624-1898-2

Designed and produced by
The Book Laboratory® Inc.
Bolinas, California

Project editor: Deborah Grandinetti
Art director: Philip Dunn
Book design: Amy Ray
Illustrator: Moon*runner* Design

Text of *On the Revolutions of Heavenly Spheres* courtesy of Annapolis: St. John's Bookstore, © 1939.
Text of *Harmonies of the World* courtesy of Annapolis: St. John's Bookstore, © 1939.
Text of *Dialogues Concerning Two New Sciences* courtesy of Dover Publications.
Text of *Principia* courtesy of New York: Daniel Adee, © 1939.
Selections from *The Principle of Relativity:*
A Collection of Papers on the Special and General Theory of Relativity,
courtesy of Dover Publications.

This book may be ordered by mail from the publisher.
Please include $2.50 for postage and handling.
But try your bookstore first!

Running Press Book Publishers
125 South Twenty-second Street
Philadelphia, Pennsylvania 19103-4399

Visit us on the web!
www.runningpress.com

CONTENTS

A NOTE ON THE TEXTS

The texts in this book are excerpts from translations of the original, printed editions. We have edited for American style and consistency. Text removed from the original manuscript is indicated by a short line. Here are other relevant details:

On the Revolutions of Heavenly Spheres, by Nicolaus Copernicus, was first published in 1543 under the title *De revolutionibus orbium colestium*. This translation is by Charles Glen Wallis.

Dialogues Concerning Two New Sciences, by Galileo Galilei, was originally published in 1638 under the title *Discorsi e Dimostrazioni Matematiche, intorno à due nuoue scienze*, by the Dutch publisher Louis Elzevir. Our text is based on the translation by Henry Crew and Alfonso deSalvio.

We have selected Book Five of the five-book *Harmonies of the World* by Johannes Kepler. Kepler completed the work on May 27, 1816, publishing it under the title, *Harmonices Mundi*. This translation is by Charles Glen Wallis.

The Principia, by Isaac Newton, was originally published in 1687 under the title of *Philosophiae naturalis principia mathematica (The Mathematical Principles of Natural Philosophy)*. This translation is by Andrew Motte.

We haven chosen seven works by Albert Einstein from *The Principles of Relativity: A Collection of Original Papers on the Special Theory of Relativity*, by H.A. Lorentz, A. Einstein, H. Minkowski and H. Weyl. The entire collection was originally published in German, under the title *Des Relativitatsprinzip* in 1922. Our text comes from the translation by W. Perrett and G.B. Jeffery.

The Editors

Shown on the opposite page is the universe according to Ptolemy. One of the most influential Greek astronomers of his time (c. 165 B.C.E.), Ptolemy propounded the geocentric theory in a form that prevailed for 1400 years.

Ptolemy's view of the sun, the planets, and the stars have long been discarded, but our perceptions are still Ptolemaic. We look to the east to see the sun rise (when in relation to Earth it is stationary); we still watch the heavens move over us and use the north, south, east, west directions, ignoring the fact that our Earth is a globe.

INTRODUCTION

"If I have seen farther, it is by standing on the shoulders of giants," wrote Isaac Newton in a letter to Robert Hooke in 1676. Although Newton was referring to his discoveries in optics rather than his more important work on gravity and the laws of motion, it is an apt comment on how science, and indeed the whole of civilization, is a series of incremental advances, each building on what went before. This is the theme of this fascinating volume, which uses the original texts to trace the evolution of our picture of the heavens from the revolutionary claim of Nicolaus Copernicus that the Earth orbits the sun to the equally revolutionary proposal of Albert Einstein that space and time are curved and warped by mass and energy. It is a compelling story because both Copernicus and Einstein have brought about profound changes in what we see as our position in the order of things. Gone is our privileged place at the center of the universe, gone are eternity and certainty, and gone are Absolute Space and Time to be replaced by rubber sheets.

It is no wonder both theories encountered violent opposition: the inquisition in the case of the Copernican theory and the Nazis in the case of relativity. We now have a tendency to dismiss as primitive the earlier world picture of Aristotle and Ptolemy in which the Earth was at the center and the Sun went round it. However we should not be too scornful of their model, which was anything but simpleminded. It incorporated Aristotle's deduction that the earth is a round ball rather than a flat plate, and it was reasonably accurate in its main function, that of predicting the apparent positions of the heavenly bodies in the sky for astrological purposes. In fact, it was about as accurate as the heretical suggestion put forward in 1543 by Copernicus that the Earth and the planets moved in circular orbits around the Sun.

Galileo found Copernicus' proposal convincing not because it better fit the observations of planetary positions but because of its simplicity and elegance, in contrast to the complicated epicycles of the Ptolemaic model. In *Dialogues Concerning Two New Sciences*, Galileo's characters, Salviati and Sagredo, put forward persuasive arguments in support of

Copernicus. Yet it was still possible for his third character, Simplicio, to defend Aristotle and Ptolemy and to maintain that in reality the earth was at rest and the sun went round the earth.

It was not until Kepler's work made the Sun-centered model more accurate and Newton gave it laws of motion that the Earth-centered picture finally lost all credibility. It was quite a shift in our view of the universe: If we are not at the center, is our existence of any importance? Why should God or the Laws of Nature care about what happens on the third rock from the sun, which is where Copernicus has left us? Modern scientists have out-Copernicused Copernicus by seeking an account of the universe in which Man (in the old prepolitically correct sense) played no role. Although this approach has succeeded in finding objective impersonal laws that govern the universe, it has not (so far at least) explained why the universe is the way it is rather than being one of the many other possible universes that would also be consistent with the laws.

Some scientists would claim that this failure is only provisional, that when we find the ultimate unified theory, it will uniquely prescribe the state of the universe, the strength of gravity, the mass and charge of the electron and so on. However, many features of the universe (like the fact that we are on the third rock, rather than the second or fourth) seem arbitrary and accidental and not the predictions of a master equation. Many people (myself included) feel that the appearance of such a complex and structured universe from the simple laws requires the invocation of something called the anthropic principle, which restores us to the central position we have been too modest to claim since the time of Copernicus. The anthropic principle is based on the self-evident fact that we wouldn't be asking questions about the nature of the universe if the universe hadn't contained stars, planets and stable chemical compounds, among other prerequisites of (intelligent?) life as we know it. If the ultimate theory made a unique prediction for the state of the universe and its contents, it would be a remarkable coincidence that this state was in the small subset that allows life.

However, the work of the last thinker in this volume, Albert Einstein, raises a new possibility. Einstein played an important role in the development of quantum theory which says that a system doesn't just have a single history as one might have thought. Rather it has every possible history with some probability. Einstein was also almost solely responsible for the general theory of relativity in which space and time are curved and become dynamic. This means that they are subject to quantum theory and that the universe itself has every possible shape and history. Most of these histories will be quite unsuitable for the development of life, but a very few have all the conditions needed. It doesn't matter if these few have a very low probability relative to the others: The lifeless universes will have no one to observe them. It is sufficient that there is at least one history in which life develops, and we ourselves are evidence for that, though maybe not for intelligence. Newton said he was "*standing on the shoulders of giants.*" But as this volume illustrates so well, our understanding doesn't advance just by slow and steady building on previous work. Sometimes as with Copernicus and Einstein, we have to make the intellectual leap to a new world picture. Maybe Newton should have said, "*I used the shoulders of giants as a springboard.*"

Nicolaus Copernicus (1473-1543)

HIS LIFE AND WORK

Nicolaus Copernicus, a sixteenth-century Polish priest and mathematician, is often referred to as the founder of modern astronomy. That credit goes to him because he was the first to conclude that the planets and Sun did not revolve around the Earth. Certainly there was speculation that a heliocentric—or Sun-centered—universe had existed as far back as Aristarchus (d. 230 B.C.E.), but the idea was not seriously considered before Copernicus. Yet to understand the contributions of Copernicus, it is important to consider the religious and cultural implications of scientific discovery in his time.

As far back as the fourth century B.C.E., the Greek thinker and philosopher Aristotle (384–322 B.C.E.) devised a planetary system in his book, *On the Heavens*, (*De Caelo*) and concluded that because the Earth's shadow on the Moon during eclipses was always round, the world was spherical in shape rather than flat. He also surmised the Earth was round because when one watched a ship sail out to sea one noticed that the hull disappeared over the horizon before the sails did.

In Aristotle's geocentric vision, the Earth was stationary and the planets Mercury, Venus, Mars, Jupiter, and Saturn, as well as the Sun and the Moon performed circular orbits around the Earth. Aristotle also believed the stars were fixed to the celestial sphere, and his scale of the universe purported these fixed stars to be not much further beyond the orbit of Saturn. He believed in perfect circular motions and had good evidence to believe the Earth to be at rest. A stone dropped from a tower fell straight down. It did not fall to the west, as we would expect it to do if the Earth rotated from west to east. (Aristotle did not consider that the stone might partake in the Earth's rotation). In an attempt to combine physics with the metaphysical, Aristotle devised his theory of a "prime mover," which held that a mystical force behind the fixed stars caused the circular motions he observed. This model of the universe was accepted and embraced by theologians, who often interpreted prime movers as angels, and Aristotle's vision endured for centuries. Many modern scholars believe universal acceptance of this theory by religious authorities hindered the progress of science, as to challenge Aristotle's theories was to call into question the authority of the Church itself.

Ptolemy's geocentric model of the universe.

Five centuries after Aristotle's death, an Egyptian named Claudius Ptolemaeus (Ptolemy, 87–150 C.E.), created a model for the universe that more accurately predicted the movements and actions of spheres in the heavens. Like Aristotle, Ptolemy believed the Earth was stationary. Objects fell to the center of the Earth, he reasoned, because the Earth must be fixed at the center of the universe. Ptolemy ultimately elaborated a system in which the celestial bodies moved around the circumference of their own epicycles (a circle in which a planet moves and which has a center that is itself carried around at the same time on the circumference of a larger circle. To accomplish this, he put the Earth slightly off center of the universe and called this new center the "equant"—an imaginary point that helped him account for observable planetary movements. By custom designing the sizes of circles, Ptolemy was better able to predict the motions of celestial bodies. Western Christendom had little quarrel with Ptolemy's geocentric system, which left room in the universe behind the fixed stars to accommodate a heaven and a hell, and so the Church adopted the Ptolemaic model of the universe as truth.

Aristotle and Ptolemy's picture of the cosmos reigned, with few significant modifications, for well over a thousand years. It wasn't until 1514 that the Polish priest Nicolaus Copernicus revived the heliocentric model of the universe. Copernicus proposed it merely as a model for calculating planetary positions, because he was concerned that the Church might label him a heretic if he proposed it as a description of reality. Copernicus became convinced, through his own study of planetary motions, that the Earth was merely another planet and the Sun was the center of the universe. This hypothesis became known as a heliocentric model. Copernicus' breakthrough marked one of the greatest paradigm shifts in world history, opening the way to modern astronomy and broadly affecting science, philosophy, and religion. The elderly priest was hesitant to divulge his theory, lest it provoke Church authorities to any angry response, and so he withheld his work from all but a few astronomers. Copernicus' landmark *De Revolutionibus* was published while he was on his deathbed, in 1543. He did not live long enough to witness the chaos his heliocentric theory would cause.

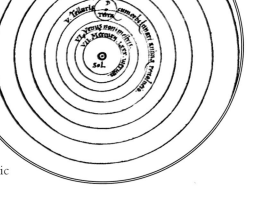

Copernicus' heliocentric model of the universe.

Copernicus was born on February 19, 1473 in Torun, Poland, into a family of merchants and municipal officials who placed a high priority on education. His uncle, Lukasz Watzenrode, prince-bishop of Ermland, ensured that his nephew received the best academic training available in Poland. In 1491, Copernicus enrolled at Cracow University, where he pursued a course of general studies for four years before traveling to Italy to study law and medicine, as was common practice among Polish elites at the time. While studying at the University of Bologna (where he would eventually become a professor of astronomy), Copernicus boarded at the home of Domenico Maria de Novara, the renowned mathematician of whom Copernicus would ultimately become a disciple. Novara was a critic of Ptolemy, whose second-century astronomy he regarded with skepticism. In November 1500, Copernicus observed a lunar eclipse in Rome. Although he spent the next few years in Italy studying medicine, he never lost his passion for astronomy.

A lunar eclipse in 1500 first stimulated Copernicus' interest in astronomy.

After receiving the degree of Doctor of Canon Law, Copernicus practiced medicine at the episcopal court of Heilsberg, where his uncle lived. Royalty and high clergy requested his medical services, but Copernicus spent most of his time in service of the poor. In 1503, he returned to Poland and moved into his uncle's bishopric palace in Lidzbark Warminski. There he tended to the administrative matters of the diocese, as well as serving as an advisor to his uncle. After his uncle's death in 1512, Copernicus moved permanently to Frauenburg and would spend the rest of his life in priestly service. But the man who was a scholar in mathematics, medicine, and theology was only beginning the work for which he would become best known.

In March of 1513, Copernicus purchased 800 building stones and a barrel of lime from his chapter so that he could build an observation tower. There, he made use of astronomical instruments such as quadrants, parallactics, and astrolabes to observe the sun, moon, and stars. The following year, he wrote a brief *Commentary on the Theories of the Motions of Heavenly Objects from Their Arrangements (De hypothesibus motuum coelestium a se constitutis commentariolus)*, but he refused to publish the manuscript and only discreetly circulated it among his most trusted friends. *The Commentary* was a first attempt to propound an astronomical theory that the Earth moves and the Sun remains at rest. Copernicus had become dissatisfied with the Aristotelian-Ptolemaic astronomical system that had dominated Western thought for centuries. The center of the Earth, he thought, was not the center of the universe, but merely the center of the Moon's orbit. Copernicus had come to believe that apparent perturbations in the observable motion of the planets was a result of the Earth's own rotation around its axis and of its travel in orbit. "We revolve around the Sun," he concluded in *Commentary*, "like any other planet.

Despite speculation about a Sun-centered universe as far back as the third century B.C.E. by Aristarchus, theologians and intellectuals felt more comfortable with a geocentric theory, and the premise was barely challenged in earnest. Copernicus prudently abstained from disclosing any of his views in public, preferring to develop his ideas quietly by exploring mathematical calculations and drawing elaborate diagrams, and to keep his theories from circulating outside of a select group of friends. When, in 1514, Pope Leo X summoned Bishop Paul of Fossombrone to recruit Copernicus to offer an opinion on reforming the ecclesiastical calendar, the Polish astronomer replied that knowledge of the motions of the Sun and Moon in relation to the length of the year was insufficient to have any bearing on reform. The challenge must have preoccupied Copernicus, however, for he later wrote to Pope Paul III, the same pope who commissioned Michaelangelo to paint the Sistine Chapel, with some relevant observations, which later served to form the foundation of the Gregorian calendar seventy years later.

Still, Copernicus feared exposing himself to the contempt of the populace and the Church, and he spent years working privately to amend and expand the *Commentary*. The result was *On the Revolutions of Heavenly Spheres (De Revolutionibus Orbium Coelestium)* which he completed in 1530, but withheld from publication for thirteen years. The risk of the Church's condemnation was not, however, the only reason for Copernicus' hesitancy to publish. Copernicus was a perfectionist and considered his observations in constant need of verification and revision.

PTOLEMY USING AN ASTROLABE

Ptolemy was often confused with the Egyptian kings, so he is shown wearing a crown.

Theology and Astronomy in discourse. The Church expected theories of astronomy to be consistent with official doctrines of theology.

He continued to lecture on these principles of his planetary theory, even appearing before Pope Clement VII, who approved of his work. In 1536, Clement formally requested that Copernicus publish his theories. But it took a former pupil, 25-year-old Georg Joachim Rheticus of Germany, who relinquished his chair in mathematics in Wittenberg so that he could study under Copernicus, to persuade his master to publish *On the Revolutions*. In 1540, Rheticus assisted in the editing of the work and presented the manuscript to a Lutheran printer in Nuremberg, ultimately giving birth to the Copernican Revolution.

When *On the Revolutions* appeared in 1543, it was attacked by Protestant theologians who held the premise of a heliocentric universe to be unbiblical. Copernicus' theories, they reasoned, might lead people to believe that they are simply part of a natural order, and not the masters of nature, the center around which nature was ordered. Because of this clerical opposition, and perhaps also general incredulity at the prospect of a non-geocentric universe, between 1543 and 1600, fewer than a dozen scientists embraced Copernican theory. Still, Copernicus had done nothing to resolve the major problem facing any system in which the Earth rotated on its axis (and revolved around the Sun), namely, how it is that terrestrial bodies stay with the rotating Earth. The answer was proposed by Giordano Bruno, an Italian scientist and avowed Copernican, who suggested that space might have no boundaries and that the solar system might be one of many such systems in the universe. Bruno also expanded on some purely speculative areas of astronomy that Copernicus did not explore in *On the Revolutions*. In his writings and lectures, the Italian scientist held that there were infinite worlds in the universe with intelligent life, some perhaps with beings superior to humans. Such audacity brought Bruno to the attention of the Inquisition, which tried and condemned him for his heretical beliefs. He was burned at the stake in 1600.

On the whole, however, the book did not have an immediate impact on modern astronomic study. In *On the Revolutions*, Copernicus did not actually put forth a heliocentric system, but rather a heliostatic one. He considered the Sun to be not precisely at the center of the universe, but

only close to it, so as to account for variations in observable retrogression and brightness. The Earth, he asserted, made one full rotation on its axis daily, and orbited around the Sun once yearly. In the first section of the book's six sections, he took issue with the Ptolemaic system, which placed all heavenly bodies in orbit around the Earth, and established the correct heliocentric order: Mercury, Venus, Mars, Jupiter, and Saturn (the six planets known at the time). In the second section, Copernicus used mathematics (namely epicycles and equants) to explain the motions of the stars and planets, and reasoned that the Sun's motion coincided with that of the Earth. The third section gives a mathematical explanation of the precession of the equinoxes, which Copernicus attributes to the Earth's gyration around its axis. The remaining sections of *On the Revolutions* focus on the motions of the planets and the Moon.

People condemned by the Inquisition were burned.

Copernicus holding a model
of his heliocentric theory
of the universe.

Copernicus was the first to position Venus and Mercury correctly, establishing with remarkable accuracy the order and distance of the known planets. He saw these two planets (Venus and Mercury) as being closer to the Sun, and noticed that they revolved at a faster rate inside the Earth's orbit.

Before Copernicus, the Sun was thought to be another planet. Placing the Sun at the virtual center of the planetary system was the beginning of the Copernican revolution. By moving the Earth away from the center of the universe, where it was presumed to anchor all heavenly bodies, Copernicus was forced to address theories of gravity. Pre-Copernican gravitational explanations had posited a single center of gravity (the Earth), but Copernicus theorized that each heavenly body might have its own gravitational qualities and asserted that heavy objects everywhere tended toward their own center. This insight would eventually lead to the theory of universal gravitation, but its impact was not immediate.

By 1543, Copernicus had become paralyzed on his right side and weakened both physically and mentally. The man who was clearly a perfectionist had no choice but to surrender control of his manuscript, *On the Revolutions*, in the last stages of printing. He entrusted his student, George Rheticus with the manuscript, but when Rheticus was forced to leave Nuremberg, the manuscript fell into the hands of Lutheran theologian Andreas Osiander. Osiander, hoping to appease advocates of the geocentric theory, made several alterations without Copernicus's knowledge and consent. Osiander placed the word "hypothesis" on the title page, deleted important passages, and added his own sentences which diluted the impact and certainty of the work. Copernicus was said to have received a copy of the printed book in Frauenburg on his deathbed, unaware of Osiander's revisions. His ideas lingered in relative obscurity for nearly one hundred years, but the seventeenth century would see men like Galileo Galilei, Johannes Kepler, and Isaac Newton build on his theories of a heliocentric universe, effectively obliterating Aristotelian ideas. Many have written about the unassuming Polish priest who would change the way people saw the universe, but the German writer and scientist Johann

Wolfgang von Goethe may have been the most eloquent when he wrote of the contributions of Copernicus:

Of all discoveries and opinions, none may have exerted a greater effect on the human spirit than the doctrine of Copernicus. The world had scarcely become known as round and complete in itself when it was asked to waive the tremendous privilege of being the center of the universe. Never, perhaps, was a greater demand made on mankind—for by this admission so many things vanished in mist and smoke! What became of Eden, our world of innocence, piety and poetry; the testimony of the senses; the conviction of a poetic-religious faith? No wonder his contemporaries did not wish to let all this go and offered every possible resistance to a doctrine which in its converts authorized and demanded a freedom of view and greatness of thought so far unknown, indeed not even dreamed of.

—Johann Wolfgang von Goethe

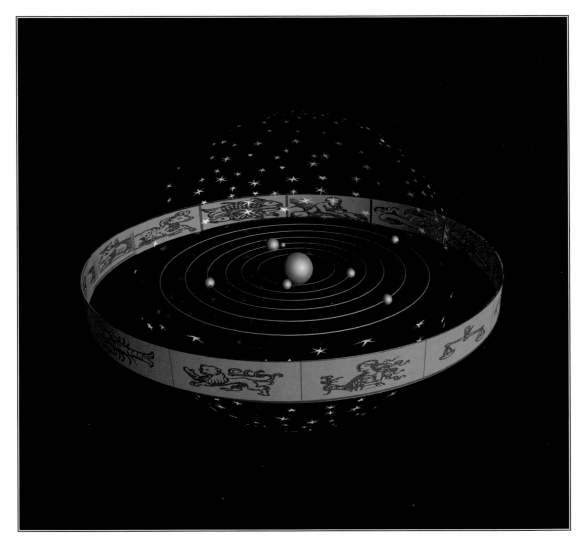

THE UNIVERSE ACCORDING TO COPERNICUS WITH THE ASTROLOGICAL LINK

For those who studied the "heavens" astronomy and astrology were the same thing.
They were also called The Celestial Sciences.

ON THE REVOLUTION OF THE HEAVENLY SPHERES

INTRODUCTION TO THE READER CONCERNING THE HYPOTHESIS OF THIS WORK[1]

Since the newness of the hypotheses of this work—which sets the Earth in motion and puts an immovable Sun at the center of the universe—has already received a great deal of publicity, I have no doubt that certain of the savants have taken grave offense and think it wrong to raise any disturbance among liberal disciplines which have had the right set-up for a long time now. If, however, they are willing to weigh the matter scrupulously, they will find that the author of this work has done nothing which merits blame. For it is the job of the astronomer to use painstaking and skilled observation in gathering together the history of the celestial movements, and then since he cannot by any line of reasoning reach the true causes of these movements—to think up or construct whatever causes or hypotheses he pleases such that, by the assumption of these causes, those same movements can be calculated from the principles of geometry for the past and for the future too. This artist is markedly outstanding in both of these respects: for it is not necessary that these hypotheses should be true, or even probably; but it is enough if they provide a calculus which fits the observations—unless by some chance there is anyone so ignorant of geometry and optics as to hold the epicycle of Venus as probable and to believe this to be a cause why Venus alternately precedes and follows the Sun at an angular distance of up to 40° or more. For who does not see that it necessarily follows from this assumption that the diameter of the planet in its perigee should appear more than four times greater, and the body of the planet more than sixteen times greater, than in its apogee? Nevertheless the experience of all the ages is opposed to that.[2] There are also other things in this discipline which are just as absurd, but it is not necessary to examine them right now. For it is sufficiently clear that this art is absolutely and profoundly ignorant of the causes of the apparent irregular movements. And if it constructs and thinks up causes—and it has certainly thought up a good many—nevertheless it does not think them up in order to persuade anyone of their truth but only in order that they may provide a correct basis for calculation. But since for one and the same movement varying hypotheses are proposed from time to time, as eccentricity or epicycle for the movement of the Sun, the astronomer much prefers to take the one which is easiest to grasp. Maybe the philosopher demands probability instead; but neither of them will grasp anything certain or hand it on, unless it has been divinely revealed to him. Therefore let us permit these new hypotheses to make a public appearance among old ones which are themselves no more probable, especially since they are wonderful and easy and bring with them a vast storehouse of learned observations. And as far as hypotheses go, let no one expect anything in the way of certainty from astronomy, since astronomy can offer us nothing certain, lest, if anyone take as true that which has been constructed for another use, he go away from this discipline a bigger fool than when he came to it. Farewell.

————

BOOK ONE[3]

Among the many and varied literary and artistic studies upon which the natural talents of man are nourished, I think that those above all should be embraced and pursued with the most loving care which have to do with things that are very beautiful and very worthy of knowledge. Such studies are those which deal with the godlike circular movements of the world and the course of the stars, their magnitudes, distances, risings, and settings, and the causes of the other appearances in the heavens; and which finally explicate the whole form. For what could be more beautiful than the heavens which contain all beautiful things? Their very names make this clear: *Caelum* (heavens) by naming that which is beautifully carved; and *Mundus* (world), purity and elegance. Many philosophers have called the world a visible god on account of its extraordinary excellence. So if the worth of the arts were measured by the matter with which they deal, this art—which some call astronomy, others astrology, and many of the ancients the consummation of mathematics—would be by far the most outstanding. This art which is as it were the head of all the liberal arts and the one most worthy of a free man leans upon nearly all the other branches of mathematics. Arithmetic, geometry, optics,

geodesy, mechanics, and whatever others, all offer themselves in its service. And since a property of all good arts is to draw the mind of man away from the vices and direct it to better things, these arts can do that more plentifully, over and above the unbelievable pleasure of mind (which they furnish). For who, after applying himself to things which he sees established in the best order and directed by divine ruling, would not through diligent contemplation of them and through a certain habituation be

awakened to that which is best and would not wonder at the Artificer of all things, in Whom is all happiness and every good? For the divine Psalmist surely did not say gratuitously that he took pleasure in the workings of God and rejoiced in the works of His hands, unless by means of these things as by some sort of vehicle we are transported to the contemplation of the highest Good.

Now as regards the utility and ornament which they confer upon a commonwealth—to pass over the innumerable advantages they give to private citizens—Plato makes an extremely good point, for in the seventh book of the *Laws* he says that this study should be pursued in especial, that through it the orderly arrangement of days into months and years and the determination of the times for solemnities and sacrifices should keep the state alive and watchful; and he says that if anyone denies that this study is necessary for a man who is going to take up any of the highest branches of learning, then such a person is thinking foolishly; and he thinks that it is impossible for anyone to become godlike or be called so who has no necessary knowledge of the Sun, Moon, and the other stars.

However, this more divine than human science, which inquires into the highest things, is not lacking in difficulties. And in particular we see that as regards its principles and assumptions, which the Greeks call "hypotheses," many of those who undertook to deal with them were not in accord and hence did not employ the same methods of calculation. In addition, the courses of the planets and the revolution of the stars cannot be determined by exact calculations and reduced to perfect knowledge unless, through the passage of time and with the help of many prior observations, they can, so to speak, be handed down to posterity. For even if Claud Ptolemy of Alexandria, who stands far in front of all the others on account of his wonderful care and industry, with the help of more than forty years of observations brought this art to such a high point that there seemed to be nothing left which he had not touched upon; nevertheless we see that very many things are not in accord with the movements which should follow from his doctrine but rather with movements which were discovered later and were unknown to him. Whence even Plutarch in speaking of the revolving solar year says, "So far

OPPOSITE PAGE

A rendering of the solar system as we see it today, very much a confirmation of what Copernicus envisioned.

Peter Apian's sixteenth-century proof that the Earth is spherical.

the movement of the stars has overcome the ingenuity of the mathematicians." Now to take the year itself as my example, I believe it is well known how many different opinions there are about it, so that many people have given up hope of risking an exact determination of it. Similarly, in the case of the other planets I shall try—with the help of God, without Whom we can do nothing—to make a more detailed inquiry concerning them, since the greater the interval of time between us and the founders of this art—whose discoveries we can compare with the new ones made by us—the more means we have of supporting our own theory. Furthermore, I confess that I shall expound many things differently from my predecessors—although with their aid, for it was they who first opened the road of inquiry into these things.

I. THE WORLD IS SPHERICAL

In the beginning we should remark that the world is globe-shaped; whether because this figure is the most perfect of all, as it is an integral whole and needs no joints; or because this figure is the one having the greatest volume and thus is especially suitable for that which is going to comprehend and conserve all things; or even because the separate parts of the world *i.e.*, the Sun, Moon, and stars are viewed under such a form; or because everything in the world tends to be delimited by this form, as is apparent in the case of drops of water and other liquid bodies, when they become delimited of themselves. And so no one would hesitate to say that this form belongs to the heavenly bodies.

2. THE EARTH IS SPHERICAL TOO

The Earth is globe-shaped too, since on every side it rests upon its center. But it is not perceived straightway to be a perfect sphere, on account of the great height of its mountains and the lowness of its valleys, though they modify its universal roundness to only a very small extent.

That is made clear in this way. For when people journey northward from anywhere, the northern vertex of the axis of daily revolution gradually moves overhead, and the other moves downward to the same extent; and many stars situated to the north are seen not to set, and many to the south are seen not to rise any more. So Italy does not see Canopus, which is visible to Egypt. And Italy sees the last star of Fluvius, which is not visible to this region situated in a more frigid zone. Conversely, for people who travel southward, the second group of stars becomes higher in the sky; while those become lower which for us are high up.

Moreover, the inclinations of the poles have everywhere the same ratio with places at equal distances from the poles of the Earth and that happens in no other figure except the spherical. Whence it is manifest that the Earth itself is contained between the vertices and is therefore a globe.

Add to this the fact that the inhabitants of the East do not perceive the evening eclipses of the Sun and Moon; nor the inhabitants of the West, the morning eclipses; while of those who live in the middle region—some see them earlier and some later.

Furthermore, voyagers perceive that the waters too are fixed within this figure; for example, when land is not visible from the deck of a ship, it may be seen from the top of the mast, and conversely, if something shining is attached to the top of the mast, it appears to those remaining on the shore to come down gradually, as the ship moves from the land, until finally it becomes hidden, as if setting.

Moreover, it is admitted that water, which by its nature flows, always seeks lower places—the same way as earth—and does not climb up the shore any farther than the convexity of the shore allows. That is why the land is so much higher where it rises up from the ocean.

The Earth from space, showing how land and water make up a single globe.

3. HOW LAND AND WATER MAKE UP A SINGLE GLOBE

And so the ocean encircling the land pours forth its waters everywhere and fills up the deeper hollows with them. Accordingly it was necessary for there to be less water than land, so as not to have the whole Earth soaked with water—since both of them tend toward the same center on account of their weight—and so as to leave some portions of land—such as the islands discernible here and there—for the preservation of living creatures. For what is the continent itself and the *orbis terrarum* except an island which is larger than the rest? We should not listen to certain Peripatetics who maintain that there is ten times more water than land and who arrive at that conclusion because in the transmutation of the elements the liquefaction of one part of Earth results in ten parts of water. And they say that land has emerged for a certain distance because, having hollow spaces inside, it does not balance everywhere with respect to weight and so the center of gravity is different from the center of magnitude. But they fall into error through ignorance of geometry; for they do not know that there cannot be seven times more water than land and some part of the land still remain dry, unless the land abandon its center of gravity and give place to the waters as being heavier. For spheres are to one another as the cubes of their diameters. If therefore there were seven parts of water and one part of land, the diameter of the land could not be greater than the radius of the globe of the waters. So it is even less possible that the water should be ten times greater. It can be gathered that there is no difference between the centers of magnitude and of gravity of the Earth from the fact that the convexity of the land spreading out from the ocean does not swell continuously, for in that case it would repulse the sea-waters as much as possible and would not in any way allow interior seas and huge gulfs to break through. Moreover, from the seashore outward the depth of the abyss would not stop increasing, and so no island or reef or any spot of land would be met with by people voyaging out very far. Now it is well known that there is not quite the distance of two miles—at practically the center of the *orbis terrarum* between the Egyptian and the Red Sea. And on the contrary, Ptolemy in his *Cosmography* extends inhabitable lands as far as the median circle, and he leaves that part of the Earth as unknown, where the moderns have added Cathay and other vast regions as far as 60° longitude, so that inhabited land extends in longitude farther than the rest of the ocean does. And if you add to these the islands discovered in our time under the princes of Spain and Portugal and especially America—named after the ship's captain who discovered her—which they consider a second *orbis terrarum* on account of her so far unmeasured magnitude—besides many other islands heretofore unknown, we would not be greatly surprised if there were antiphodes or antichthones. For reasons of geometry compel us to believe that America is situated diametrically opposite to the India of the Ganges.

And from all that I think it is manifest that the land and the water rest upon one center of gravity; that this is the same as the center of magnitude of the land, since land is the heavier; that parts of land which are as it were yawning are filled with water; and that accordingly there is little water in comparison with the land, even if more of the surface appears to be covered by water.

Copernicus' mapping of the land and water was remarkably accurate for his time.

Now it is necessary that the land and the surrounding waters have the figure which the shadow of the Earth casts, for it eclipses the Moon by projecting a perfect circle upon it. Therefore the Earth is not a plane, as Empedocles and Anaximenes opined; or a tympanoid, as Leucippus; or a scaphoid, as Heracleitus; or hollowed out in any other way, as Democritus; or again a cylinder, as Anaximander; and it is not infinite in its lower part, with the density increasing rootwards, as Xenophanes thought; but it is perfectly round, as the philosophers perceived.

4. THE MOVEMENT OF THE CELESTIAL BODIES IS REGULAR, CIRCULAR, AND EVERLASTING—OR ELSE COMPOUNDED OF CIRCULAR MOVEMENTS

After this we will recall that the movement of the celestial bodies is circular. For the motion of a sphere is to turn in a circle; by this very act expressing its form, in the most simple body, where beginning and end cannot be discovered or distinguished from one another, while it moves through the same parts in itself.

But there are many movements on account of the multitude of spheres or orbital circles.[4] The most obvious of all is the daily revolution—which the Greeks call νυχθήμερν *i.e.*, having the temporal span of a day and a night. By means of this movement the whole world—with the exception of the Earth—is supposed to be borne from east to west. This movement is taken as the common measure of all movements, since we measure even time itself principally by the number of days.

Next, we see other as it were antagonistic revolutions; *i.e.*, from west to east, on the part of the Sun, Moon, and the wandering stars. In this way the Sun gives us the year, the Moon the months—the most common periods of time; and each of the other five planets follows its own cycle. Nevertheless these movements are manifoldly different from the first movement. First, in that they do not revolve around the same poles as the first movement but follow the oblique ecliptic; next, in that they do not seem to move in their circuit regularly. For the Sun and Moon are caught moving at times more slowly and at times more quickly. And we perceive the five wandering stars sometimes even to retrograde and to come to a stop between these two movements. And though the Sun always proceeds straight ahead along its route, they wander in various ways, straying sometimes towards the south, and at other times towards the north—whence they are called "planets." Add to this the fact that sometimes they are nearer the Earth—and are then said to be at their perigee—and at other times are farther away—and are said to be at their apogee.

We must however confess that these movements are circular or are composed of many circular movements, in that they maintain these irregularities in accordance with a constant law and with fixed periodic returns: and that could not take place, if they were not circular. For it is only the circle which can bring back what is past and over with; and in this way, for example, the sun by a movement composed of circular movements brings back to us the inequality of days and nights and the four seasons of the year. Many movements are recognized in that movement, since it is impossible that a simple heavenly body should be moved irregularly by a single sphere. For that would have to take place either on account of the inconstancy of the motor virtue—whether by reason of an extrinsic cause or its intrinsic nature—or on account of the inequality between it and the moved body. But since the mind shudders at either of these suppositions, and since it is quite unfitting to suppose that such a state of affairs exists among things which are established in the best system, it is agreed that their regular movements appear to us as irregular, whether on account of their circles having different poles or even because the Earth is not at the center of the circles in which they revolve. And so for us watching from the Earth, it happens that the transits of the planets, on account of being at unequal distances from the Earth, appear greater when they are nearer than when they are farther away, as has been shown in optics: Thus in the case of equal arcs of an orbital circle which are seen at different distances there will appear to be unequal movements in equal times. For this reason I think it necessary above all that we should note carefully what the relation of the Earth to the heavens is, so as not—when we wish to scrutinize the highest things—to be ignorant of those which are nearest to us, and so as not—by the same error—to attribute to the celestial bodies what belongs to the Earth.

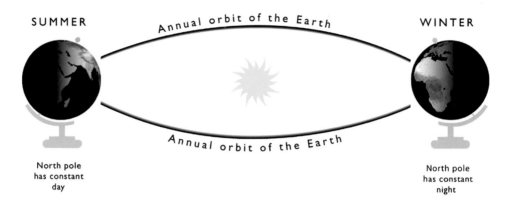

SUMMER

Annual orbit of the Earth

WINTER

Annual orbit of the Earth

North pole
has constant
day

North pole
has constant
night

5. DOES THE EARTH HAVE A CIRCULAR MOVEMENT? AND OF ITS PLACE

Now that it has been shown that the Earth too has the form of a globe, I think we must see whether or not a movement follows upon its form and what the place of the Earth is in the universe. For without doing that it will not be possible to find a sure reason for the movements appearing in the heavens. Although there are so many authorities for saying that the Earth rests in the center of the world that people think the contrary supposition inopinable and even ridiculous; if however we consider the thing attentively, we will see that the question has not yet been decided and accordingly is by no means to be scorned. For every apparent change in place occurs on account of the movement either of the thing seen or of the spectator, or on account of the necessarily unequal movement of both. For no movement is perceptible relatively to things moved equally in the same directions—I mean relatively to the thing seen and the spectator. Now it is from the Earth that the celestial circuit is beheld and presented to our sight. Therefore, if some movement should belong to the Earth it will appear, in the parts of the universe which are outside, as the same movement but in the opposite direction, as though the things outside were passing over. And the daily revolution in especial is such a movement. For the daily revolution appears to carry the whole universe along, with the exception of the Earth and the things around it.

And if you admit that the heavens possess none of this movement but that the Earth turns from west to east, you will find—if you make a serious examination—that as regards the apparent rising and setting of the sun, moon, and stars the case is so. And since it is the heavens which contain and embrace all things as the place common to the universe, it will not be clear at once why movement should not be assigned to the contained rather than to the container, to the thing placed rather than to the thing providing the place.

As a matter of fact, the Pythagoreans Herakleides and Ekphantus were of this opinion and so was Hicetas the Syracusan in Cicero; they made the Earth to revolve at the center of the world. For they believed that the stars set by reason of the interposition of the Earth and that with cessation of that they rose again. Now upon this assumption there follow other things, and a no smaller problem concerning the place of the Earth, though it is taken for granted and believed by nearly all that the Earth is the center of the world. For if anyone denies that the Earth occupies the midpoint or center of the world yet does not admit that the distance (between the two) is great enough to be compared with (the distance to) the sphere of the fixed stars but is considerable and quite apparent in relation to the orbital circles of the Sun and the planets; and if for that reason he thought that their movements appeared irregular because they are organized around a different center from the center of the Earth, he might perhaps be able to bring forward a perfectly sound reason for movement which appears irregular. For the fact that the wandering stars are seen to be sometimes nearer the Earth and at other times farther away necessarily argues that the center of the Earth is not the center of their circles. It is not yet clear whether the Earth draws near to them and moves away or they draw near to the Earth and move away.

And so it would not be very surprising if someone attributed some other movement to the earth in addition to the daily revolution. As a matter of fact, Philolaus the Pythagorean—no ordinary mathematician, whom Plato's biographers say Plato went to Italy for the sake of seeing— is supposed to have held that the Earth moved in a circle and wandered in some other movements and was one of the planets.

OPPOSITE PAGE

Copernicus' explanation of a planetary loop.

Many however have believed that they could show by geometrical reasoning that the Earth is in the middle of the world; that it has the proportionality of a point in relation to the immensity of the heavens, occupies the central position, and for this reason is immovable, because, when the universe moves, the center remains unmoved and the things which are closest to the center are moved the most slowly.

6. ON THE IMMENSITY OF THE HEAVENS IN RELATION TO THE MAGNITUDE OF THE EARTH

It can be understood that this great mass which is the Earth is not comparable with the magnitude of the heavens, from the fact that the boundary circles—for that is the translation of the Greek ορίζοντες— cut the whole celestial sphere into two halves; for that could not take place if the magnitude of the Earth in comparison with the heavens, or its distance from the center of the world, were considerable. For the circle bisecting a sphere goes through the center of the sphere, and is the greatest circle which it is possible to circumscribe.

Now let the horizon be the circle *ABCD*, and let the Earth, where our point of view is, be *E*, the center of the horizon by which the visible stars are separated from those which are not visible. Now with a dioptra or horoscope or level placed at *E*, the beginning of Cancer is seen to rise at point *C*; and at the same moment the beginning of Capricorn appears to set at *A*. Therefore, since *AEC* is in a straight line with the dioptra, it is clear that this line is a diameter of the ecliptic, because the six signs bound a semicircle, whose center *E* is the same as that of the horizon. But when a revolution has taken place and the beginning of Capricorn arises at *B*, then the setting of Cancer will be visible at *D*, and *BED* will be a straight line and a diameter of the ecliptic. But it has already been seen that the line *AEC* is a diameter of the same circle; therefore, at their common section, point *E* will be their center. So in this way the horizon always bisects the ecliptic, which is a great circle of the sphere. But on a sphere, if a circle bisects one of the great circles, then the circle bisecting is a great circle. Therefore the horizon is a great circle; and its center is the same as that of the ecliptic, as far as appearance goes;

ABOVE

The Hubble space telescope has revealed that Copernicus was right about the immensity of the heavens.

OPPOSITE

His contemporaries held a contrasting view, symbolically depicted here. Atlas is shown holding the entire universe, which consists of our solar system.

A sixteenth-century Flemish armillary sphere showed a geocentric model with seven nesting planetary rings.

although nevertheless the line passing through the center of the Earth and the line touching to the surface are necessarily different; but on account of their immensity in comparison with the Earth they are like parallel lines, which on account of the great distance between the termini appear to be one line, when the space contained between them is in no perceptible ratio to their length, as has been shown in optics.

From this argument it is certainly clear enough that the heavens are immense in comparison with the Earth and present the aspect of an infinite magnitude, and that in the judgment of sense-perception the Earth is to the heavens as a point to a body and as a finite to an infinite magnitude. But we see that nothing more than that has been shown, and it does not follow that the Earth must rest at the center of the world. And we should be even more surprised if such a vast world should wheel completely around during the space of twenty-four hours rather than that its least part, the Earth, should. For saying that the center is immovable and that those things which are closest to the center are moved least does not argue that the Earth rests at the center of the world. That is no different from saying that the heavens revolve but the poles are at rest and those things which are closest to the poles are moved least. In this way Cynosura (the pole star) is seen to move much more slowly than Aquila or Canicula because, being very near to the pole, it describes a smaller circle, since they are all on a single sphere, the movement of which stops at its axis and which does not allow any of its parts to have movements which are equal to one another. And nevertheless the revolution of the whole brings them round in equal times but not over equal spaces.

The argument which maintains that the Earth, as a part of the celestial sphere and as sharing in the same form and movement, moves very little because very near to its center advances to the following position: therefore the Earth will move, as being a body and not a center, and will describe in the same time arcs similar to, but smaller than, the arcs of the celestical circle. It is clearer than daylight how false that is; for there would necessarily always be noon at one place and midnight at another, and so the daily risings and settings could not take place, since the movement of the whole and the part would be one and inseparable.

But the ratio between things separated by diversity of nature is so entirely different that those which describe a smaller circle turn more quickly than those which describe a greater circle. In this way Saturn, the highest of the wandering stars, completes its revolution in thirty years, and the moon which is without doubt the closest to the Earth completes its circuit in a month, and finally the Earth itself will be considered to complete a circular movement in the space of a day and a night. So this same problem concerning the daily revolution comes up again. And also the question about the place of the Earth becomes even less certain on account of what was just said. For that demonstration proves nothing except that the heavens are of an indefinite magnitude with respect to the Earth. But it is not at all clear how far this immensity stretches out. On the contrary, since the minimal and indivisible corpuscles, which are called atoms, are not perceptible to sense, they do not, when taken in twos or in some small number, constitute a visible body; but they can be taken in such a large quantity that there will at last be enough to form a visible magnitude. So it is as regards the place of the earth; for although it is not at the center of the world, nevertheless the distance is as nothing, particularly in comparison with the sphere of the fixed stars.

THE ILLUSTRATED ON THE SHOULDERS OF GIANTS

7. WHY THE ANCIENTS THOUGHT THAT THE EARTH WAS AT REST AT THE MIDDLE OF THE WORLD AS ITS CENTER

Wherefore for other reasons the ancient philosophers have tried to affirm that the Earth is at rest at the middle of the world, and as principal cause they put forward heaviness and lightness. For Earth is the heaviest element; and all things of any weight are borne towards it and strive to move towards the very center of it.

For since the Earth is a globe towards which from every direction heavy things by their own nature are borne at right angles to its surface, the heavy things would fall on one another at the center if they were not held back at the surface; since a straight line making right angles with a plane surface where it touches a sphere leads to the center. And those things which are borne toward the center seem to follow along in order to be at rest at the center. All the more then will the Earth be at rest at the center; and, as being the receptacle for falling bodies, it will remain immovable because of its weight.

They strive similarly to prove this by reason of movement and its nature. For Aristotle says that the movement of a body which is one and simple is simple, and the simple movements are the rectilinear and the circular. And of rectilinear movements, one is upward, and the other is downward. As a consequence, every simple movement is either toward the center, *i.e.*, downward, or away from the center, *i.e.*, upward, or around the center, *i.e.*, circular. Now it belongs to earth and water, which are considered heavy, to be borne downward, *i.e.*, to seek the center: for air and fire, which are endowed with lightness, move upward, *i.e.*, away from the center. It seems fitting to grant rectilinear movement to these four elements and to give the heavenly bodies a circular movement around the center—so Aristotle. Therefore, said Ptolemy of Alexandria, if the Earth moved, even if only by its daily rotation, the contrary of what was said above would necessarily take place. For this movement which would traverse the total circuit of the Earth in twenty-four hours would necessarily be very headlong and of an unsurpassable velocity. Now things which are suddenly and violently whirled around are seen to be utterly unfitted for reuniting, and the more unified are seen to become dispersed, unless some constant force constrains them to stick together. And a long time ago, he says, the scattered Earth would have passed beyond the heavens, as is certainly ridiculous; and *a fortiori* so would all the living creatures and all the other separate masses which could by no means remain unshaken. Moreover, freely falling bodies would not arrive at the places appointed them, and certainly not along the perpendicular line which they assume so quickly. And we would see clouds and other things floating in the air always borne toward the west.

8. ANSWER TO THE AFORESAID REASONS AND THEIR INADEQUACY

For these and similar reasons they say that the Earth remains at rest at the middle of the world and that there is no doubt about this. But if someone opines that the Earth revolves, he will also say that the movement is natural and not violent. Now things which are according to nature produce effects contrary to those which are violent. For things to which force or violence is applied get broken up and are unable to subsist for a long time. But things which are caused by nature are in a right condition and are kept in their best organization. Therefore Ptolemy had no reason to fear that the Earth and all things on the Earth would be scattered in a revolution caused by the efficacy of nature, which is greatly different from that of art or from that which can result from the genius of man. But why didn't he feel anxiety about the world instead, whose movement must necessarily be of greater velocity, the greater the heavens are than the Earth? Or have the heavens become so immense, because an unspeakably vehement motion has pulled them away from the center, and because the heavens would fall if they came to rest anywhere else?

Surely if this reasoning were tenable, the magnitude of the heavens would extend infinitely. For the farther the movement is borne upward by the vehement force, the faster will the movement be, on account of the ever-increasing circumference which must be traversed every twenty-four hours: and conversely, the immensity of the sky would increase with the increase in movement. In this way, the velocity would make the magnitude increase infinitely, and the magnitude the velocity. And in accordance with the axiom of physics *that that which is infinite cannot be traversed or moved in any way, then the heavens will necessarily come to rest.*

But they say that beyond the heavens there isn't any body or place or void or anything at all; and accordingly it is not possible for the heavens to move outward; in that case it is rather surprising that something can be held together by nothing. But if the heavens were infinite and were finite only with respect to a hollow space inside, then it will be said with more truth that there is nothing outside the heavens, since anything which occupied any space would be in them; but the heavens will remain immobile. For movement is the most powerful reason wherewith they try to conclude that the universe is finite.

But let us leave to the philosophers of nature the dispute as to whether the world is finite or infinite, and let us hold as certain that the Earth is held together between its two poles and terminates in a spherical surface. Why therefore should we hesitate any longer to grant to it the movement which accords naturally with its form, rather than put the whole world in a commotion—the world whose limits we do not and cannot know? And why not admit that the appearance of daily revolution belongs to the heavens but the reality belongs to the Earth? And things are as when Aeneas said in Virgil: "We sail out of the harbor, and the land and the cities move away." As a matter of fact, when a ship floats on over a tranquil sea, all the things outside seem to the voyagers to be moving in a movement which is the image of their own, and they think on the contrary that they themselves and all the things with them are at rest. So it can easily happen in the case of the movement of the Earth that the whole world should be believed to be moving in a circle. Then what would we say about the clouds and the other things floating in the air or falling or rising up, except that not only the Earth and the watery element with which it is conjoined are moved in this way but also no small part of the air and whatever other things have a similar kinship with the Earth? whether because the neighboring air, which is mixed with earthly and watery matter, obeys the same nature as the Earth or because the movement of the air is an acquired one, in which it participates without resistance on account of the contiguity and perpetual rotation of the Earth.

Conversely, it is no less astonishing for them to say that the highest region of the air follows the celestial movement, as is shown by those stars which appear suddenly—I mean those called "comets" or "bearded stars" by the Greeks. For that place is assigned for their generation; and like all the other stars they rise and set. We can say that that part of the air is deprived of terrestrial motion on account of its great distance from the Earth. Hence the air which is nearest to the Earth and the things floating in it will appear tranquil, unless they are driven to and fro by the wind or some other force, as happens. For how is the wind in the air different from a current in the sea?

But we must confess that in comparison with the world the movement of falling and of rising bodies is twofold and is in general compounded of the rectilinear and the circular. As regards things which move downward on account of their weight because they have very much earth in them, doubtless their parts possess the same nature as the whole, and it is for the same reason that fiery bodies are drawn upward with force. For even this earthly fire feeds principally on earthly matter; and they define flame as glowing smoke. Now it is a property of fire to make that which it invades to expand; and it does this with such force that it can be stopped by no means or contrivance from breaking prison and completing its job. Now expanding movement moves away from the center to the circumference; and so if some part of the Earth caught on fire, it would be borne away from the center and upward. Accordingly, as they say, a simple body possesses a simple movement—this is first verified in the case of circular movement—as long as the simple body remain in its unity in its natural place. In this place, in fact, its movement is none other than the circular, which remains entirely in itself, as though at rest. Rectilinear movement, however, is added to those bodies which journey away from their natural place or are shoved out of it or are outside it somehow. But nothing is more repugnant to the order of the whole and to the form of the world than for anything to be outside of its place. Therefore rectilinear movement belongs only to bodies which are not in the right condition and are not perfectly conformed to their nature—when they are separated from their whole and abandon its unity. Furthermore, bodies which

Compasses from the time of Copernicus.

are moved upward or downward do not possess a simple, uniform, and regular movement—even without taking into account circular movement. For they cannot be in equilibrium with their lightness or their force of weight. And those which fall downward possess a slow movement at the beginning but increase their velocity as they fall. And conversely we note that this earthly fire—and we have experience of no other—when carried high up immediately dies down, as if through the acknowledged agency of the violence of earthly matter.

Now circular movement always goes on regularly, for it has an unfailing cause; but (in rectilinear movement) the acceleration stops, because, when the bodies have reached their own place, they are no longer heavy or light, and so the movement ends. Therefore, since circular movement belongs to wholes and rectilinear to parts, we can say that the circular movement stands with the rectilinear, as does animal with sick. And the fact that Aristotle divided simple movement into three genera: away from the center, toward the center, and around the center, will be considered merely as an act of reason, just as we distinguish between line, point, and surface, though none of them can subsist without the others or without body.

In addition, there is the fact that the state of immobility is regarded as more noble and godlike than that of change and instability, which for that reason should belong to the Earth rather than to the world. I add that it seems rather absurd to ascribe movement to the container or to that which provides the place and not rather to that which is contained and has a place, *i.e.*, the Earth. And lastly, since it is clear that the wandering stars are sometimes nearer and sometimes farther away from the Earth, then the movement of one and the same body around the center—and they mean the center of the Earth—will be both away from the center and toward the center. Therefore it is necessary that movement around the center should be taken more generally; and it should be enough if each movement is in accord with its own center. You see therefore that for all these reasons it is more probably that the Earth moves than that it is at rest—especially in the case of the daily revolution, as it is the Earth's very own. And I think that is enough as regards the first part of the question.

9. WHETHER MANY MOVEMENTS CAN BE ATTRIBUTED TO THE EARTH, AND CONCERNING THE CENTER OF THE WORLD

Therefore, since nothing hinders the mobility of the Earth, I think we should now see whether more than one movement belongs to it, so that it can be regarded as one of the wandering stars. For the apparent irregular movement of the planets and their variable distances from the Earth—which cannot be understood as occurring in circles homocentric with the Earth—make it clear that the Earth is not the center of their circular movements. Therefore, since there are many centers, it is not foolhardy to doubt whether the center of gravity of the Earth rather than some other is the center of the world. I myself think that gravity or heaviness is nothing except a certain natural appetency implanted in the parts by the divine providence of the universal Artisan, in order that they should unite with one another in their oneness and wholeness and come together in the form of a globe. It is believable that this affect is present in the sun, moon, and the other bright planets and that through its efficacy they remain in the spherical figure in which they are visible, though they nevertheless accomplish their circular movements in many different ways. Therefore if the Earth too possesses movements different from the one around its center, then they will necessarily be movements which similarly appear on the outside in the many bodies; and we find the yearly revolution among these movements. For if the annual revolution were changed from being solar to being terrestrial, and immobility were granted to the sun, the risings and settings of the signs and of the fixed stars—whereby they become morning or evening stars—will appear in the same way; and it will be seen that the stoppings, retrogressions, and progressions of the wandering stars are not their own, but are a movement of the Earth and that they borrow the appearances of this movement. Lastly, the Sun will be regarded as occupying the center of the world. And the ratio of order in which these bodies succeed one another and the harmony of the whole world teaches us their truth, if only—as they say—we would look at the thing with both eyes.

———

Earthrise over the Moon.

II. A DEMONSTRATION OF THE THREEFOLD MOVEMENT OF THE EARTH

Therefore since so much and such great testimony on the part of the planets is consonant with the mobility of the Earth, we shall now give a summary of its movement, insofar as the appearances can be shown forth by its movement as by an hypothesis. We must allow a threefold movement altogether.

The first—which we said the Greeks called νυχθημὲρινος—is the proper circuit of day and night, which goes around the axis of the Earth from west to east—as the world is held to move in the opposite direction—and describes the equator or the equinoctial circle—which some, imitating the Greek expression ὶσηὲρινος, call the equidial.

The second is the annual movement of the center, which describes the circle of the (zodiacal) signs around the Sun similarly from west to east, *i.e.*, towards the signs which follow (from Aries to Taurus) and moves along between Venus and Mars, as we said, together with the bodies accompanying it. So it happens that the sun itself seems to traverse the ecliptic with a similar movement. In this way, for example, when the center of the Earth is traversing Capricorn, the Sun seems to be crossing Cancer; and when Aquarius, Leo, and so on, as we were saying.

It has to be understood that the equator and the axis of the Earth have a variable inclination with the circle and the plane of the ecliptic. For if they remained fixed and only followed the movement of the center simply, no inequality of days and nights would be apparent, but it would always be the summer solstice or the winter solstice or the equinox, or summer or winter, or some other season of the year always remaining the same. There follows then the third movement, which is the declination: it is also an annual revolution but one towards the signs which precede (from Aries to Pisces), or westwards, *i.e.*, turning back counter to the movement of the center; and as a consequence of these two movements which are nearly equal to one another but in opposite directions, it follows that the axis of the Earth and the greatest of the parallel circles on it, the equator, always look towards approximately the same quarter of the world, just as if they remained immobile. The Sun in the meanwhile is seen to move along the oblique ecliptic with that movement with which the center of the Earth moves, just as if the center of the Earth were the center of the world—provided you remember that the distance between the sun and the Earth in comparison with the sphere of the fixed stars is imperceptible to us.

Since these things are such that they need to be presented to sight rather than merely to be talked about, let us draw the circle *ABCD*, which will represent the annual circuit of the center of the Earth in the plane of the ecliptic, and let *E* be the sun around its center. I will cut this circle into four equal parts by means of the diameters *AEC* and *BED*. Let the point *A* be the beginning of Cancer; *B* of Libra; *E* of Capricorn; and *D* of Aries. Now let us put the center of the Earth first at *A*, around which we shall describe the terrestrial equator *FGHI*, but not in the same plane (as the ecliptic) except that the diameter *GAI* is the common section of the circles, *i.e.*, of the equator and the ecliptic. Also let the diameter *FAH* be drawn at right angles to *GAI*; and let *F* be the limit of the greatest southward declination (of the equator), and *H* of the northward declination. With this set-up, the Earth-dweller will see the Sun—which is at the center *E*—at the point of the winter solstice in Capricorn—which is caused by the greatest northward declination at *H* being turned toward the Sun; since the inclination of the equator with respect to line *AE* describes by means of the daily revolution the winter tropic, which is parallel to the equator at the distance comprehended by the angle of inclination *EAH*. Now let the center of the Earth proceed from west to east; and let *F*, the limit of greatest declination, have just as great a movement from east to west, until at *B* both of them have traversed quadrants of circles. Meanwhile, on account of the equality of the revolutions, angle *EAI* will always remain equal to angle *AEB*; the diameters will always stay parallel to one another—*FAH* to *FBH* and *GAI* to *GBI*; and the equator will remain parallel to the equator. And by reason of the cause spoken of many times already, these lines will appear in the immensity of the sky as the same. Therefore from the point *B* the beginning of Libra, *E* will appear to be in Aries, and the common section of the two circles (of the ecliptic and the equator) will fall upon line *GBIE*, in respect to which the daily revolution has no declination; but every declination will be on one side or the other of this line. And so the Sun will be soon in the spring equinox. Let the center of the Earth advance under the same conditions; and when it has completed a semicircle at *C*, the Sun will appear to be entering Cancer. But since *F* the southward declination of the equator is now turned toward the sun, the result is that the Sun is seen in the north, traversing the summer tropic in accordance with angle of inclination *ECF*. Again, when *F* moves on through the third quadrant of the circle, the common section *GI* will fall on line *ED*, whence the Sun, seen in Libra, will appear to have reached the autumn equinox. But then as, in the same progressive movement, *HF* gradually turns in the direction of the Sun, it will make the situation at the beginning return, which was our point of departure.

In another way: Again in the underlying plane let *AEC* be both the diameter (of the ecliptic) and its common section with the circle perpendicular to its plane. In this circle let *DGFI*, the meridian passing through the poles of the Earth be described around *A* and *C*, in turn, *i.e.*, in Cancer and in Capricorn. And let the axis of the Earth be *DF*, the north pole *D*, the south pole *F*, and *GI* the diameter of the equator. Therefore when F is turned in the direction of the Sun, which is at *E*, and the inclination of the equator is northward in proportion to angle *IAE*, then the movement around the axis will describe—with the diameter *KL* and at the distance *LI*—parallel to the equator the southern circle, which appears with respect to the Sun as the tropic of Capricorn. Or—to speak more correctly—this movement around the axis describes, in the direction of *AE,* a conic surface, which has the center of the Earth as its vertex and a circle parallel to the equator as its base.1 Moreover in the opposite sign, *C,* the same things take place but conversely. Therefore it is clear how the two mutually opposing movements, *i.e.*, that of the center and that of the inclination, force the axis of the Earth to remain balanced in the same way and to keep a similar position, and how they make all things appear as if they were movements of the Sun.

Now we said that the yearly revolutions of the center and of the declination were approximately equal, because if they were exactly so, then the points of equinox and solstice and the obliquity of the ecliptic in relation to the sphere of the fixed stars could not change at all. But as the difference is very slight, it is not revealed except as it increases with time: As a matter of fact, from the time of Ptolemy to ours there has been a precession of the equinoxes and solstices of about 21°. For that reason some have believed that the sphere of the fixed stars was moving, and so they choose a ninth higher sphere. And when that was not enough, the moderns added a tenth, but without attaining the end which we hope we shall attain by means of the movement of the Earth. We shall use this movement as a principle and a hypothesis in demonstrating other things.

———

Galileo Galilei (1564-1642)

HIS LIFE AND WORK

In 1633, ninety years after the death of Copernicus, the Italian astronomer and mathematician Galileo Galilei was taken to Rome to stand trial before the Inquisition for heresy. The charge stemmed from the publication of Galileo's *Dialogue Concerning the Two Chief World Systems: Ptolemaic and Copernican* (*Dialogo sopra Ii due massimi sistemi del mondo: Ttolemaico, e Ccopernicono*). In this book, Galileo forcefully asserted, in defiance of a 1616 edict against the propagation of Copernican doctrine, that the heliocentric system was not just a hypothesis but was the truth. The outcome of the trial was never in doubt. Galileo admitted that he might have gone too far in his arguments for the Copernican system, despite previous warnings by the Roman Catholic Church. A majority of the cardinals in the tribunal found him "vehemently suspected of heresy" for supporting and teaching the idea that the Earth moves and is not the center of the universe, and they sentenced him to life imprisonment.

Galileo was also forced to sign a handwritten confession and to renounce his beliefs publicly. On his knees, and with his hands on the Bible, he pronounced this abjuration in Latin:

I, Galileo Galilei, son of the late Vincenzio Galilei of Florence, aged 70 years, tried personally by this court, and kneeling before You, the most Eminent and Reverend Lord Cardinals, Inquisitors-General throughout the Christian Republic against heretical depravity, having before my eyes the Most Holy Gospels, and laying on them my own hands; I swear that I have always believed, I believe now,

and with God's help I will in future believe all which the Holy Catholic and Apostolic Church doth hold, preach, and teach.

But since I, after having been admonished by this Holy Office entirely to abandon the false opinion that the Sun was the center of the universe and immoveable, and that the Earth was not the center of the same and that it moved, and that I was neither to hold, defend, nor teach in any manner whatever, either orally or in writing, the said false doctrine; and after having received a notification that the said doctrine is contrary to Holy Writ, I did write and cause to be printed a book in which I treat of the said already condemned doctrine, and bring forward arguments of much efficacy in its favor, without arriving at any solution: I have been judged vehemently suspected of heresy, that is, of having held and believed that the Sun is the center of the universe and immoveable, and that the Earth is not the center of the same, and that it does move.

Nevertheless, wishing to remove from the minds of your Eminences and all faithful Christians this vehement suspicion reasonably conceived against me, I abjure with sincere heart and unfeigned faith, I curse and detest the said errors and heresies, and generally all and every error and sect contrary to the Holy Catholic Church. And I swear that for the future I will neither say nor assert in speaking or writing such things as may bring upon me similar suspicion; and if I know any heretic, or one suspected of heresy, I will denounce him to this Holy Office, or to the Inquisitor and Ordinary of the place in which I may be.

I also swear and promise to adopt and observe entirely all the penances which have been or may be by this Holy Office imposed on me. And if I contravene any of these said promises, protests, or oaths (which God forbid!) I submit myself to all the pains and penalties which by the Sacred Canons and other Decrees general and particular are against such offenders imposed and promulgated. So help me God and the Holy Gospels, which I touch with my own hands.

I Galileo Galilei aforesaid have abjured, sworn, and promised, and hold myself bound as above; and in token of the truth, with my own hand have subscribed the present schedule of my abjuration, and have recited it word by word. In Rome, at the Convent della Minerva, this 22nd day of June, 1633. I, Galileo Galilei, have abjured as above, with my own hand.

Legend has it that as Galileo rose to his feet, he uttered under his breath, "*Eppur si muove*"—"And yet, it moves." The remark captivated scientists and scholars for centuries, as it represented defiance of obscurantism and nobility of purpose in the search for truth under the most adverse circumstances. Although an oil portrait of Galileo dating from 1640 has been discovered bearing the inscription "*Eppur si muove*," most historians regard the story as myth. Still, it is entirely within Galileo's character to have only paid lip service to the Church's demands in his abjuration and then to have returned to his scientific studies, whether they adhered to non-Copernican principles or not. After all, what had brought Galileo before the Inquisition was his publication of *Two Chief World Systems*, a direct challenge to the Church's 1616 edict forbidding him from teaching the Copernican theory of the Earth in motion around the Sun as anything but a hypothesis. "*Eppur si muove*" may not have concluded his trial and abjuration, but the phrase certainly punctuated Galileo's life and accomplishments.

Galileo at his trial.

Born in Pisa on February 18, 1564, Galileo Galilei was the son of Vincenzo Galilei, a musician and mathematician. The family moved to Florence when Galileo was young, and there he began his education in a monastery. Although from an early age Galileo demonstrated a penchant for mathematics and mechanical pursuits, his father was adamant that he enter a more useful field, and so in 1581 Galileo enrolled in the University of Pisa to study medicine and the philosophy of Aristotle. It was in Pisa that Galileo's rebelliousness emerged. He had little or no interest in medicine and began to study mathematics with a passion. It is believed that while observing the oscillations of a hanging lamp in the cathedral of Pisa, Galileo discovered the isochronism of the pendulum— the period of swing is independent of its amplitude—which he would apply a half-century later in building an astronomical clock.

Galileo persuaded his father to allow him to leave the university without a degree, and he returned to Florence to study and teach mathematics. By 1586, he had begun to question the science and philosophy of Aristotle, preferring to reexamine the work of the great mathematician Archimedes, who was also known for discovering and perfecting

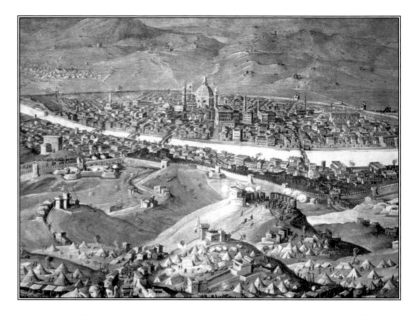

Painting of Florence at the time Galileo lived there by Giorgio Vasari.

methods of integration for calculating areas and volumes. Archimedes also gained a reputation for his invention of many machines ultimately used as engines of war, such as giant catapults to hurl boulders at an advancing army and large cranes to topple ships. Galileo was inspired mainly by Archimedes' mathematical genius, but he too was swept up in the spirit of invention, designing a hydrostatic balance to determine an object's density when weighed in water.

In 1589, Galileo became a professor of mathematics at the University of Pisa, where he was required to teach Ptolemaic astronomy—the theory that the Sun and the planets revolve around the Earth. It was in Pisa, at the age of twenty-five, that Galileo obtained a deeper understanding of astronomy and began to break with Aristotle and Ptolemy. Lecture notes recovered from this period show that Galileo had adopted the Archimedean approach to motion; specifically, he was teaching that the density of a falling object, not its weight, as Aristotle had maintained, was proportional to the speed at which it fell. Galileo is said to have demonstrated his theory by dropping objects of the different weights but the same density from atop the leaning tower of Pisa. In Pisa, too, he wrote *On motion* (*De motu*), a book that contradicted the Aristotelian theories of motion and established Galileo as a leader in scientific reformation.

The University of Padua, where Galileo made many of his discoveries.

After his father's death in 1592, Galileo did not see much of a future for himself in Pisa. The pay was dismal, and with the help of a family friend, Guidobaldo del Monte, Galileo was appointed to the chair in mathematics at the University of Padua, in the Venetian Republic. There, Galileo's reputation blossomed. He remained at Padua for eighteen years, lecturing on geometry and astronomy as well as giving private lessons on cosmography, optics, arithmetic, and the use of the sector in military engineering. In 1593, he assembled treatises on fortifications and mechanics for his private students and invented a pump that could raise water under power of a single horse.

In 1597, Galileo invented a calculating compass that proved useful to mechanical engineers and military men. He also began a correspondence with Johannes Kepler, whose book *Mystery of the Cosmos* (*Mysterium cosmographicum*) Galileo had read. Galileo sympathized with Kepler's Copernican views, and Kepler hoped that Galileo would openly support the theory of a heliocentric Earth. But Galileo's scientific interests were still focused on mechanical theories, and he did not follow Kepler's wishes. Also at that time Galileo had developed a personal interest in Marina Gamba, a Venetian woman by whom he had a son and two daughters. The eldest daughter, Virginia, born in 1600, maintained a very

THE ILLUSTRATED ON THE SHOULDERS OF GIANTS

CENTER

Chandra X-Ray Observatory image of a supernova such as the one observed above Padua in 1604.

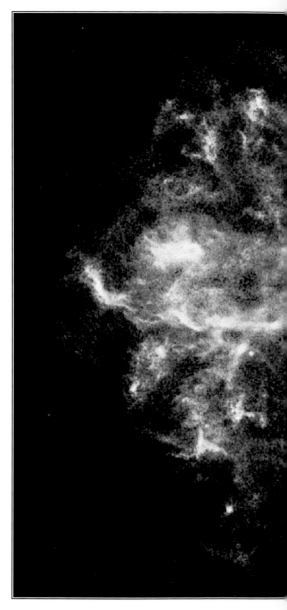

close relationship with her father, mainly through an exchange of correspondence, for she spent most of her short adult life in a convent, taking the name Maria Celeste in tribute to her father's interest in celestial matters.

In the first years of the seventeenth century, Galileo experimented with the pendulum and explored its association with the phenomenon of natural acceleration. He also began work on a mathematical model describing the motion of falling bodies, which he studied by measuring the time it took balls to roll various distances down inclined planes. In 1604, a supernova observed in the night sky above Padua renewed questions about Aristotle's model of the unchanging heavens. Galileo thrust himself into the forefront of the debate, delivering several provocative lectures, but he was hesitant to publish his theories. In October 1608, a Dutchman by the name of Hans Lipperhey applied for a patent on a spyglass that could make faraway objects appear closer. Upon hearing of the invention, Galileo set about attempting to improve it. Soon he had designed a nine-power telescope, three times more powerful than Lipperhey's device, and within a year, he

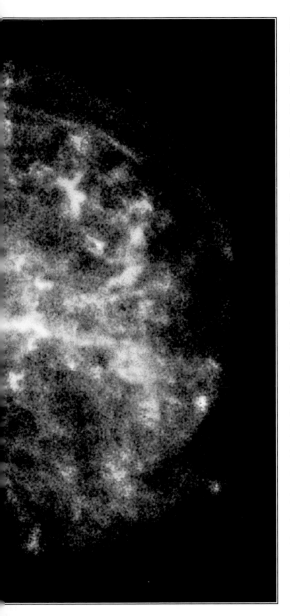

had produced a thirty-power telescope. When he pointed the scope toward the skies in January 1610, the heavens literally opened up to humankind. The Moon no longer appeared to be a perfectly smooth disc but was seen to be a mountainous and full of craters. Through his telescope, Galileo determined that the Milky Way was actually a vast gathering of separate stars. But most important, he sighted four moons around Jupiter, a discovery that had tremendous implications for many of the geocentrically inclined, who held that all heavenly bodies revolved exclusively around the Earth. That same year, he published *The Starry Messenger* (*Sidereus Nuncius*), in which he announced his discoveries and which put him in the forefront of contemporary astronomy. He felt unable to continue teaching Aristotelian theories, and his renown enabled him to take a position in Florence as mathematician and philosopher to the grand duke of Tuscany.

Once free from the responsibilities of teaching, Galileo was able to devote himself to telescopy. He soon observed the phases of Venus, which confirmed Copernicus' theory that the planet revolved around the Sun.

He also noted Saturn 's oblong shape, which he attributed to numerous moons revolving around the planet, for his telescope was unable to detect Saturn's rings.

The Roman Catholic Church affirmed and praised Galileo's discoveries but did not agree with his interpretations of them. In 1613, Galileo published *Letters on Sunspots,* marking the first time in print that he had defended the Copernican system of a heliocentric universe. The work was immediately attacked and its author denounced, and the Holy Inquisition soon took notice. When in 1616 Galileo published a theory of tides, which he believed was proof that the Earth moved, he was summoned to Rome to answer for his views. A council of theologians issued an edict that Galileo was practicing bad science when he taught the Copernican system as fact. But Galileo was never officially condemned. A meeting with Pope Paul V led him to believe that the pontiff held him in esteem and that he could continue to lecture under the pontiff 's protection. He was, however, strongly warned that Copernican theories ran contrary to the Scriptures and that they may only be presented as hypotheses.

When upon Paul's death in 1623 one of Galileo's friends and supporters, Cardinal Barberini, was elected pope, taking the name Urban VIII, Galileo presumed that the 1616 edict would be reversed. Urban told Galileo that he himself was responsible for omitting the word "heresy" from the edict and that as long as Galileo treated Copernican doctrine as hypothesis and not truth, he would be free to publish. With this assurance, over the next six years Galileo worked on *Dialogue Concerning the Two Chief World Systems*, the book that would lead to his imprisonment.

Two Chief World Systems takes the form of a polemic between an advocate of Aristotle and Ptolemy and a supporter of Copernicus, who seek to win an educated everyman over to the respective philosophies. Galileo prefaced the book with a statement in support of the 1616 edict against him, and by presenting the theories through the book's characters, he is able to avoid openly declaring his allegiance to either side. The public clearly perceived, nonetheless, that in *Two Chief World Systems* Galileo was disparaging Aristotelianism. In the polemic,

Aristotle's cosmology is only weakly defended by its simpleminded supporter and is viciously attacked by the forceful and persuasive Copernican. The book achieved a great success, despite being the subject of massive protest upon publication. By writing it in vernacular Italian rather than Latin, Galileo made it accessible to a broad range of literate Italians, not just to churchmen and scholars. Galileo's Ptolemaic rivals were furious at the dismissive treatment that their scientific views had been given. In Simplicio, the defender of the Ptolemaic system, many readers recognized a caricature of Simplicius, a sixth-century Aristotelian commentator. Pope Urban VIII, meanwhile, thought that Simplicio was meant as a caricature of himself. He felt misled by Galileo, who apparently had neglected to inform him of any injunction in the 1616 edict when he sought permission to write the book. Galileo, on the other hand, never received a written injunction, and seemed to be unaware of any violations on his part.

By March 1632, the Church had ordered the book's printer to discontinue publication, and Galileo was summoned to Rome to defend himself. Pleading serious illness, Galileo refused to travel, but the pope insisted, threatening to have Galileo removed in chains. Eleven months later, Galileo appeared in Rome for trial. He was made to abjure the heresy of the Copernican theory and was sentenced to life imprisonment. Galileo's life sentence was soon commuted to gentle house arrest in Siena under the guard of Archbishop Ascanio Piccolomini, a former student of Galileo's. Piccolomini permitted and even encouraged Galileo to resume writing. There, Galileo began his final work, *Dialogues Concerning Two New Sciences*, an examination of his accomplishments in physics. But the following year, when Rome got word of the preferential treatment Galileo was receiving from Piccolomini, it had him removed to another home, in the hills above Florence. Some historians believe that it was upon his transfer that Galileo actually said "Eppur si muove," rather than at his public abjuration following the trial.

The transfer brought Galileo closer to his daughter Virginia, but soon she died, after a brief illness, in 1634. The loss devastated Galileo, but eventually he was able to resume working on *Two New Sciences*, and he finished the book within a year. However, the Congregation of the

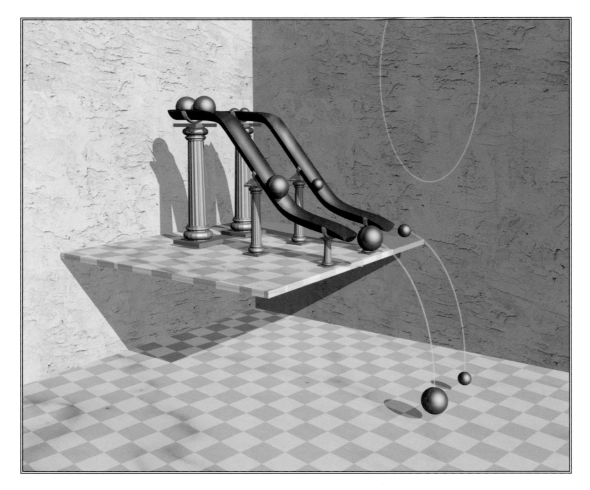

Index, the Church censor, would not allow Galileo to publish it. The manuscript had to be smuggled out of Italy to Leiden, in Protestant northern Europe, by Louis Elsevier, a Dutch publisher, before it could appear in print in 1638. *Dialogues Concerning Two New Sciences*, which set out the laws of accelerated motion governing falling bodies, is widely held to be the cornerstone of modern physics. In this book, Galileo reviewed and refined his previous studies of motion, as well as the principles of mechanics. The two new sciences Galileo focuses on are the study of the strength of materials (a branch of engineering), and the study of motion (kinematics, a branch of mathematics). In the first half of the book, Galileo described his inclined-plane experiments in accelerated motion. In the second half, Galileo took on the intractable problem of calculating the path of a projectile fired from a cannon. At first it had been thought that, in keeping with Aristotelian principles, a projectile followed a straight line until it lost its "impetus" and fell straight to the ground. Later, observers noticed that it actually returned to Earth on a curved path, but the reason this happened and an exact description of the curve no one could say—until Galileo. He concluded that the projectile's path is determined by two motions—one vertical, caused by gravity, which forces the projectile down, and one horizontal, governed by the principle of inertia.

Galileo demonstrated that the combination of these two independent motions determined the projectile's course along a mathematically describable curve. He showed this by rolling a bronze ball coated in ink down an inclined plane and onto a table, whence it fell freely off the edge and onto the floor. The inked ball left a mark on the floor where it hit, always some distance out from the table's edge. Thus Galileo proved that the ball continued to move horizontally, at a constant speed, while gravity pulled it down vertically. He found that the distance increased in proportion to the square of the time elapsed. The curve achieved a precise mathematical shape, which the ancient Greeks had termed a parabola.

So great a contribution to physics was *Two New Sciences* that scholars have long maintained that the book anticipated Isaac Newton's laws of motion. By the time of its publication, however, Galileo had gone blind. He lived out the remaining years of his life in Arcetri, where he died on January 8, 1642. Galileo's contributions to humanity were never understated. Albert Einstein recognized this when he wrote: "Propositions arrived at purely by logical means are completely empty as regards reality. Because Galileo saw this, and particularly because he drummed it into the scientific world, he is the father of modern physics—indeed of modern science."

In 1979, Pope John Paul II stated that the Roman Catholic Church may have mistakenly condemned Galileo, and he called for a commission specifically to reopen the case. Four years later, the commission reported that Galileo should not have been condemned, and the Church published all the documents relevant to his trial. In 1992, the pope endorsed the commission's conclusion.

A depiction of Galileo's telescope, the book in which he wrote the notes in this volume, the Jupiter moons on an orrery, and the planet Jupiter in the distance.

DIALOGUES CONCERNING TWO NEW SCIENCES

FIRST DAY

Interlocutors: Salviati, Sagredo and Simplicio

———

Salv. We can take wood and see it go up in fire and light, but we do not see them recombine to form wood; we see fruits and flowers and a thousand other solid bodies dissolve largely into odors, but we do not observe these fragrant atoms coming together to form fragrant solids. But where the senses fail us reason must step in; for it will enable us to understand the motion involved in the condensation of extremely rarefied and tenuous substances just as clearly as that involved in the expansion and dissolution of solids. Moreover we are trying to find out how it is possible to produce expansion and contraction in bodies which are capable of such changes without introducing vacua and without giving up the impenetrability of matter; but this does not exclude the possibility of there being materials which possess no such properties and do not, therefore, carry with them consequences which you call inconvenient and impossible. And finally, Simplicio, I have, for the sake of you philosophers, taken pains to find an explanation of how expansion and contraction can take place without our admitting the penetrability of matter and introducing vacua, properties which you deny and dislike; if you were to admit them, I should not oppose you so vigorously. Now either admit these difficulties or accept my views or suggest something better.

Sagr. I quite agree with the Peripatetic philosophers in denying the penetrability of matter. As to the vacua I should like to hear a thorough discussion of Aristotle's demonstration in which he opposes them, and what you, Salviati, have to say in reply. I beg of you, Simplicio, that you give us the precise proof of the Philosopher and that you, Salviati, give us the reply.

Simp. So far as I remember, Aristotle inveighs against the ancient view that a vacuum is a necessary prerequisite for motion and that the latter could not occur without the former. In opposition to this view Aristotle shows that it is precisely the phenomenon of motion, as we shall see, which renders untenable the idea of a vacuum. His method is to divide the argument into two parts. He first supposes bodies of different weights to move in the same medium; then supposes, one and the same body to move in different media. In the first case, he supposes bodies of different weight to move in one and the same medium with different speeds which stand to one another in the same ratio as the weights; so that, for example, a body which is ten times as heavy as another will move ten times as rapidly as the other. In the second case he assumes that the speeds of one and the same body moving in different media are in inverse ratio to the densities

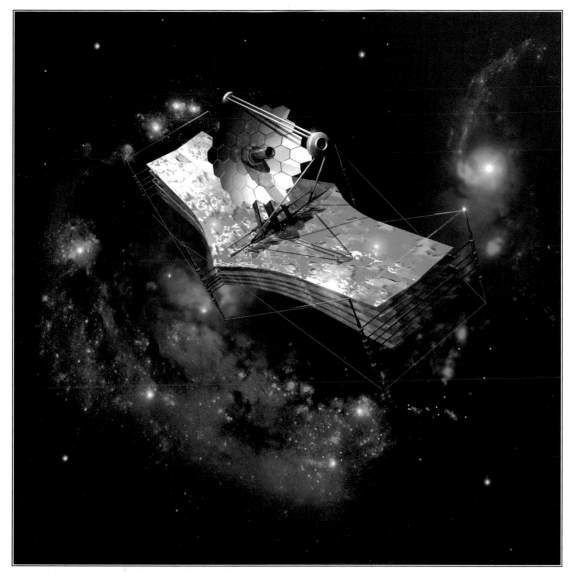

THE WEBB SPACE TELESCOPE WILL SUPERSEDE THE HUBBLE IN 2011

Galileo's entire work is completely justified by the future that is being created now. The Hubble telescope weighs over one ton, but the new Webb will be made of light hexagonal mirrors six meters across and will be 10 to 100 times more powerful than the Hubble.

of these media; thus, for instance, if the density of water were ten times that of air, the speed in air would be ten times greater than in water. From this second supposition, he shows that, since the tenuity of a vacuum differs infinitely from that of any medium filled with matter however rare, any body which moves in a plenum through a certain space in a certain time ought to move through a vacuum instantaneously; but instantaneous motion is an impossibility; it is therefore impossible that a vacuum should be produced by motion.

Salv. The argument is, as you see, *ad hominem*, that is, it is directed against those who thought the vacuum a prerequisite for motion. Now if I admit the argument to be conclusive and concede also that motion cannot take place in a vacuum, the assumption of a vacuum considered absolutely and not with reference to motion, is not thereby invalidated. But to tell you what the ancients might possibly have replied and in order to better understand just how conclusive Aristotle's demonstration is, we may, in my opinion, deny both of his assumptions. And as to the first, I greatly doubt that Aristotle ever tested by experiment whether it be true that two stones, one weighing ten times as much as the other, if allowed to fall, at the same instant, from a height of, say, 100 cubits, would so differ in speed that when the heavier had reached the ground, the other would not have fallen more than 10 cubits.

Simp. His language would seem to indicate that he had tried the experiment, because he says: *We see the heavier*; now the word *see* shows that he had made the experiment.

Sagr. But I, Simplicio, who have made the test can assure you that a cannon ball weighing one or two hundred pounds, or even more, will not reach the ground by as much as a span ahead of a musket ball weighing only half a pound, provided both are dropped from a height of 200 cubits.

Salv. But, even without further experiment, it is possible to prove clearly, by means of a short and conclusive argument, that a heavier body does not move more rapidly than a lighter one provided both bodies are of the same material and in short such as those mentioned by Aristotle. But tell me, Simplicio, whether you admit that each falling body acquires a definite speed fixed by nature, a velocity which cannot be increased or

Supposedly Galileo dropped balls of various sizes and weights off the side of the Tower of Pisa in order to find if they all fell at the same rate.

by the use of force [*violenza*] or resistance.

Simp. There can be no doubt but that one and the same body moving in a single medium has a fixed velocity which is determined by nature and which cannot be increased except by the addition of momentum [*impeto*] or diminished except by some resistance which retards it.

Salv. If then we take two bodies whose natural speeds are different, it is clear that on uniting the two, the more rapid one will be partly retarded by the slower, and the slower will be somewhat hastened by the swifter. Do you not agree with me in this opinion?

Simp. You are unquestionably right.

Salv. But if this is true, and if a large stone moves with a speed of, say, eight while a smaller moves with a speed of four, then when they are united, the system will move with a speed less than eight; but the two stones when tied together make a stone larger than that which before moved with a speed of eight. Hence the heavier body moves with less speed than the lighter; an effect which is contrary to your supposition. Thus you see how, from your assumption that the heavier body moves more rapidly than the lighter one, I infer that the heavier body moves more slowly.

Simp. I am all at sea because it appears to me that the smaller stone when added to the larger increases its weight and by adding weight I do not see how it can fail to increase its speed or, at least, not to diminish it.

Salv. Here again you are in error, Simplicio, because it is not true that the smaller stone adds weight to the larger.

Simp. This is, indeed, quite beyond my comprehension.

Salv. It will not be beyond you when I have once shown you the mistake under which you are laboring. Note that it is necessary to distinguish between heavy bodies in motion and the same bodies at rest. A large stone placed in a balance not only acquires additional weight by having another stone placed upon it, but even by the addition of a handful of hemp its weight is augmented six to ten ounces according to the quantity of hemp. But if you tie the hemp to the stone and allow them to fall freely from some height, do you believe that the hemp will press down upon the stone and thus accelerate its motion or do you think the motion will be retarded by a partial upward pressure? One always feels the pressure upon his shoulders when he prevents the motion of a load resting upon him; but if one descends just as rapidly as the load would fall how can it gravitate or press upon him? Do you not see that this would be the same as trying to strike a man with a lance when he is running away from you with a speed which is equal to, or even greater than, that with which you are following him? You must therefore conclude that, during free and natural fall, the small stone does not press upon the larger and consequently does not increase its weight as it does when at rest.

Simp. But what if we should place the larger stone upon the smaller?

Salv. Its weight would be increased if the larger stone moved more rapidly; but we have already concluded that when the small stone moves more slowly it retards to some extent the speed of the larger, so that the combination of the two, which is a heavier body than the larger of the two stones, would move less rapidly, a conclusion which is contrary to your hypothesis. We infer therefore that large and small bodies move with the same speed provided they are of the same specific gravity.

Simp. Your discussion is really admirable; yet I do not find it easy to believe that a bird-shot falls as swiftly as a cannon ball.

Salv. Why not say a grain of sand as rapidly as a grindstone? But, Simplicio, I trust you will not follow the example of many others who divert the discussion from its main intent and fasten upon some statement of mine which lacks a hair's-breadth of the truth and, under this hair, hide the fault of another which is as big as a ship's cable. Aristotle says that "an iron ball of one hundred pounds falling from a height of one hundred cubits reaches the ground before a one-pound ball has fallen a single cubit." I say that they arrive at the same time. You find, on making the experiment, that the larger outstrips the smaller by two finger-breadths, that is, when the larger has reached the ground, the other is short of it by two finger-breadths; now you would not hide behind these two fingers the ninety-nine cubits of Aristotle, nor would you mention my small error and at the same time pass over in silence his very large one. Aristotle declares that bodies of different weights, in the same medium, travel (insofar as their motion depends upon gravity) with speeds which are proportional to their weights; this he illustrates by use of bodies in which it is possible to perceive the pure and unadulterated effect of gravity, eliminating other considerations, for example, figure as being of small importance [*minimi momenti*], influences which are greatly dependent upon the medium which modifies the single effect of gravity alone. Thus we observe that gold, the densest of all substances, when beaten out into a very thin leaf, goes floating through the air; the same thing happens with stone when ground into a very fine powder. But if you wish to maintain the general proposition you will have to show that the same ratio of speeds is preserved in the case of all heavy bodies, and

Galileo's telescopes.

that a stone of twenty pounds moves ten times as rapidly as one of two; but I claim that this is false and that, if they fall from a height of fifty or a hundred cubits, they will reach the earth at the same moment.

Simp. Perhaps the result would be different if the fall took place not from a few cubits but from some thousands of cubits.

Salv. If this were what Aristotle meant you would burden him with another error which would amount to a falsehood; because, since there is no such sheer height available on Earth, it is clear that Aristotle could not have made the experiment; yet he wishes to give us the impression of his having performed it when he speaks of such an effect as one which we see.

Simp. In fact, Aristotle does not employ this principle, but uses the other one which is not, I believe, subject to these same difficulties.

Salv. But the one is as false as the other; and I am surprised that you yourself do not see the fallacy and that you do not perceive that if it were true that, in media of different densities and different resistances, such as water and air, one and the same body moved in air more rapidly than in water, in proportion as the density of water is greater than that of air, then it would follow that any body which falls through air ought also to fall through water. But this conclusion is false inasmuch as many bodies which descend in air not only do not descend in water, but actually rise.

Simp. I do not understand the necessity of your inference; and in addition I will say that Aristotle discusses only those bodies which fall in both media, not those which fall in air but rise in water.

Salv. The arguments which you advance for the Philosopher are such as he himself would have certainly avoided so as not to aggravate his first mistake. But tell me now whether the density [*corpulenza*] of the water, or whatever it may be that retards the motion, bears a definite ratio to the density of air which is less retardative; and if so fix a value for it at your pleasure.

Simp. Such a ratio does exist; let us assume it to be ten; then, for a body which falls in both these media, the speed in water will be ten times slower than in air.

Salv. I shall now take one of those bodies which fall in air but not in water, say a wooden ball, and I shall ask you to assign to it any speed you please for its descent through air.

Simp. Let us suppose it moves with a speed of twenty.

Salv. Very well. Then it is clear that this speed bears to some smaller speed the same ratio as the density of water bears to that of air; and the value of this smaller speed is two. So that really if we follow exactly the assumption of Aristotle we ought to infer that the wooden ball which falls in air, a substance ten times less-resisting than water, with a speed of twenty would fall in water with a speed of two, instead of coming to the

OPPOSITE PAGE

An astronaut dropped a lead ball and a feather in the near vacuum of the Moon and both dropped at the same rate.

surface from the bottom as it does; unless perhaps you wish to reply, which I do not believe you will, that the rising of the wood through the water is the same as its falling with a speed of two. But since the wooden ball does not go to the bottom, I think you will agree with me that we can find a ball of another material, not wood, which does fall in water with a speed of two.

Simp. Undoubtedly we can; but it must be of a substance considerably heavier than wood.

Salv. That is it exactly. But if this second ball falls in water with a speed of two, what will be its speed of descent in air? If you hold to the rule of Aristotle you must reply that it will move at the rate of twenty; but twenty is the speed which you yourself have already assigned to the wooden ball; hence this and the other heavier ball will each move through air with the same speed. But now how does the Philosopher harmonize this result with his other, namely, that bodies of different weight move through the same medium with different speeds—speeds which are proportional to their weights? But without going into the matter more deeply, how have these common and obvious properties escaped your notice?

Have you not observed that two bodies which fall in water, one with a speed a hundred times as great as that of the other, will fall in air with speeds so nearly equal that one will not surpass the other by as much as one hundredth part? Thus, for example, an egg made of marble will descend in water one hundred times more rapidly than a hen's egg, while in air falling from a height of twenty cubits the one will fall short of the other by less than four finger breadths. In short, a heavy body which sinks through ten cubits of water in three hours will traverse ten cubits of air in one or two pulse beats; and if the heavy body be a ball of lead it will easily traverse the ten cubits of water in less than double the time required for ten cubits of air. And here, I am sure, Simplicio, you find no ground for difference or objection. We conclude, therefore, that the argument does not bear against the existence of a vacuum; but if it did, it would only do away with vacua of considerable size which neither I nor, in my opinion, the ancients ever believed to exist in nature, although they might possibly be produced by force [*violenza*] as may be gathered from various experiments whose description would here occupy too much time.

Sagr. Seeing that Simplicio is silent, I will take the opportunity of saying something. Since you have clearly demonstrated that bodies of different weights do not move in one and the same medium with velocities proportional to their weights, but that they all move with the same speed, understanding of course that they are of the same substance or at least of the same specific gravity; certainly not of different specific gravities, for I hardly think you would have us believe a ball of cork moves with the same speed as one of lead; and again since you have clearly demonstrated that one and the same body moving through differently resisting media does not acquire speeds which are inversely proportional to the resistances, I am curious to learn what are the ratios actually observed in these cases.

———————

We come now to the other questions, relating to pendulums, a subject which may appear to many exceedingly arid, especially to those philosophers who are continually occupied with the more profound questions of nature. Nevertheless, the problem is one which I do not scorn. I am encouraged by the example of Aristotle whom I admire especially because he did not fail to discuss every subject which he thought in any degree worthy of consideration.

Impelled by your queries I may give you some of my ideas concerning certain problems in music, a splendid subject, upon which so many eminent men have written: among these is Aristotle himself who has discussed numerous interesting acoustical questions. Accordingly, if on the basis of some easy and tangible experiments, I shall explain some striking phenomena in the domain of sound, I trust my explanations will meet your approval.

Sagr. I shall receive them not only gratefully but eagerly. For, although I take pleasure in every kind of musical instrument and have paid considerable attention to harmony, I have never been able to fully

Pendulum in motion.

understand why some combinations of tones are more pleasing than others, or why certain combinations not only fail to please but are even highly offensive. Then there is the old problem of two stretched strings in unison; when one of them is sounded, the other begins to vibrate and to emit its note; nor do I understand the different ratios of harmony [*forme delle consonanze*] and some other details.

Salv. Let us see whether we cannot derive from the pendulum a satisfactory solution of all these difficulties. And first, as to the question whether one and the same pendulum really performs its vibrations, large, medium, and small, all in exactly the same time, I shall rely upon what I have already heard from our Academician. He has clearly shown that the time of descent is the same along all chords, whatever the arcs which subtend them, as well along an arc of 180° (*i. e.*, the whole diameter) as along one of 100°, 60°, 10°, 2°, 1/2°, or 4'. It is understood, of course, that these arcs all terminate at the lowest point of the circle, where it touches the horizontal plane.

If now we consider descent along arcs instead of their chords then, provided these do not exceed 90°, experiment shows that they are all traversed in equal times; but these times are greater for the chord than for the arc, an effect which is all the more remarkable because at first glance one would think just the opposite to be true. For since the terminal points of the two motions are the same and since the straight line included between these two points is the shortest distance between them, it would seem reasonable that motion along this line should be executed in the shortest time; but this is not the case, for the shortest time—and therefore the most rapid motion—is that employed along the arc of which this straight line is the chord.

As to the times of vibration of bodies suspended by threads of different lengths, they bear to each other the same proportion as the square roots of the lengths of the thread; or one might say the lengths are to each other as the squares of the times; so that if one wishes to make the vibration-time of one pendulum twice that of another, he must make its suspension four times as long. In like manner, if one pendulum has a suspension nine times as long as another, this second pendulum will execute three vibrations during each one of the first; from which it follows that the lengths of the suspending cords bear to each other the [inverse] ratio of the squares of the number of vibrations performed in the same time.

Sagr. Then, if I understand you correctly, I can easily measure the length of a string whose upper end is attached at any height whatever even if this end were invisible and I could see only the lower extremity. For if I attach to the lower end of this string a rather heavy weight and give it a to-and-fro motion, and if I ask a friend to count a number of its vibrations, while I, during the same time interval, count the number of vibrations of a pendulum which is exactly one cubit in length, then knowing the number of vibrations which each pendulum makes in the given interval of time one can determine the length of the string. Suppose, for example, that my friend counts 20 vibrations of the long cord during the same time in which I count 240 of my string which is one cubit in length; taking the squares of the two numbers, 20 and 240, namely 400 and 57600, then, I say, the long string contains 57600 units of such length that my pendulum will contain 400 of them; and since the length of my string is one cubit, I shall divide 57600 by 400 and thus obtain 144. Accordingly I shall call the length of the string 144 cubits.

Salv. Nor will you miss it by as much as a hand's breadth, especially if you observe a large number of vibrations.

Sagr. You give me frequent occasion to admire the wealth and profusion of nature when, from such common and even trivial phenomena, you derive facts which are not only striking and new but which are often far removed from what we would have imagined. Thousands of times I have observed vibrations especially in churches where lamps, suspended by long cords, had been inadvertently set into motion; but the most which I could infer from these observations was that the view of those who think that such vibrations are maintained by the medium is highly improbable: for, in that case, the air must needs have considerable judgment and little else to do but kill time by pushing to and fro a pendent weight with perfect regularity. But I never dreamed of learning that one and the same body, when suspended from a string a hundred cubits long and pulled aside through an arc of 90° or even 1° or 1/2°, would employ

Engraving of Galileo's pendulum clock. Galileo used his research on pendulums for a practical design.

the same time in passing through the least as through the largest of these arcs; and, indeed, it still strikes me as somewhat unlikely. Now I am waiting to hear how these same simple phenomena can furnish solutions for those acoustical problems-solutions which will be at least partly satisfactory.

Salv. First of all one must observe that each pendulum has its own time of vibration so definite and determinate that it is not possible to make it move with any other period [*altro periodo*] than that which nature has given it. For let any one take in his hand the cord to which the weight is attached and try, as much as he pleases, to increase or diminish the frequency [*frequenza*] of its vibrations; it will be time wasted. On the other hand, one can confer motion upon even a heavy pendulum which is at rest by simply blowing against it; by repeating these blasts with a frequency which is the same as that of the pendulum one can impart considerable motion. Suppose that by the first puff we have displaced the pendulum from the vertical by, say, half an inch; then if, after the pendulum has returned and is about to begin the second vibration, we add a second puff, we shall impart additional motion; and so on with other blasts provided they are applied at the right instant, and not when the pendulum is coming toward us since in this case the blast would impede rather than aid the motion. Continuing thus with many impulses [*impulsi*] we impart to the pendulum such momentum [*impeto*] that a greater impulse [*forza*] than that of a single blast will be needed to stop it.

Sagr. Even as a boy, I observed that one man alone by giving these impulses at the right instant was able to ring a bell so large that when four, or even six, men seized the rope and tried to stop it they were lifted from the ground, all of them together being unable to counterbalance the momentum which a single man, by properly-timed pulls, had given it.

Salv. Your illustration makes my meaning clear and is quite as well fitted, as what I have just said, to explain the wonderful phenomenon of the strings of the cittern [*cetera*] or of the spinet [*cimbalo*], namely, the fact that a vibrating string will set another string in motion and cause it to sound not only when the latter is in unison but even when it differs from the former by an octave or a fifth. A string which has been struck begins to vibrate and continues the motion as long as one hears the sound [*risonanza*];

these vibrations cause the immediately surrounding air to vibrate and quiver; then these ripples in the air expand far into space and strike not only all the strings of the same instrument but even those of neighboring instruments. Since that string which is tuned to unison with the one plucked is capable of vibrating with the same frequency, it acquires, at the first impulse, a slight oscillation; after receiving two, three, twenty, or more impulses, delivered at proper intervals, it finally accumulates a vibratory motion equal to that of the plucked string, as is clearly shown by equality of amplitude in their vibrations. This undulation expands through the air and sets into vibration not only strings, but also any other body which happens to have the same period as that of the plucked string. Accordingly if we attach to the side of an instrument small pieces of bristle or other flexible bodies, we shall observe that, when a spinet is sounded, only those pieces respond that have the same period as the string which has been struck; the remaining pieces do not vibrate in response to this string, nor do the former pieces respond to any other tone.

If one bows the base string on a viola rather smartly and brings near it a goblet of fine, thin glass having the same tone [*tuono*] as that of the string, this goblet will vibrate and audibly resound. That the undulations of the medium are widely dispersed about the sounding body is evinced by the fact that a glass of water may be made to emit a tone merely by the friction of the finger-tip upon the rim of the glass; for in this water is produced a series of regular waves. The same phenomenon is observed to better advantage by fixing the base of the goblet upon the bottom of a rather large vessel of water filled nearly to the edge of the goblet; for if, as before, we sound the glass by friction of the finger, me shall see ripples spreading with the utmost regularity and with high speed to large distances about the glass. I have often remarked, in thus sounding a rather large glass nearly full of water, that at first the waves are spaced with great uniformity, and when, as sometimes happens, the tone of the glass jumps an octave higher I have noted that at this moment each of the aforesaid waves divides into two; a phenomenon which shows clearly that the ratio involved in the octave [*forma dell' ottava*] is two.

Sagr. More than once have I observed this same thing, much to my

*Tuning fork in water,
showing the force of
sound vibration.*

delight and also to my profit. For a long time I have been perplexed about these different harmonies since the explanations hitherto given by those learned in music impress me as not sufficiently conclusive. They tell us that the diapason, *i.e.* the octave, involves the ratio of two, that the dia–pente which we call the fifth involves a ratio of 3:2, etc.; because if the open string of a monochord be sounded and afterwards a bridge be placed in the middle and the half length be sounded one hears the octave; and if the bridge be placed at 1/3 the length of the string, then on plucking first the open string and afterwards 2/3 of its length the fifth is given; for this reason they say that the octave depends upon the ratio of two to one [*contenuta tra'l due e l'uno*] and the fifth upon the ratio of three to two. This explanation does not impress me as sufficient to establish 2 and 3/2 as the natural ratios of the octave and the fifth; and my reason for thinking so is as follows. There are three different ways in which the tone of a string may be sharpened, namely, by shortening it, by stretching it and by making it thinner. If the tension and size of the string remain constant one obtains the octave by shortening it to one-half, i. e., by sounding first the open string and then one-half of it; but if length and size remain constant and one attempts to produce the octave by stretching he will find that it does not suffice to double the stretching weight; it must be quadrupled; so that, if the fundamental note is produced by a weight of one pound, four will be required to bring out the octave.

And finally if the length and tension remain constant, while one changes the size of the string he will find that in order to produce the octave the size must be reduced to 1/4 that which gave the fundamental. And what I have said concerning the octave, namely, that its ratio as derived from the tension and size of the string is the square of that derived from the length, applies equally well to all other musical intervals [intervalli musici].

Thus if one wishes to produce a fifth by changing the length he finds that the ratio of the lengths must be sesquialteral, in other words he sounds first the open string, then two-thirds of it; but if he wishes to produce this same result by stretching or thinning the string then it becomes necessary to square the ratio 3/2 that is by taking 9/4 [dupla sesquiquarta]; accordingly, if the fundamental requires a weight of 4 pounds, the higher note will be produced not by 6, but by 9 pounds; the same is true in regard to size, the string which gives the fundamental is larger than that which yields the fifth in the ratio of 9 to 4.

In view of these facts, I see no reason why those wise philosophers should adopt 2 rather than 4 as the ratio of the octave, or why in the case of the fifth they should employ the sesquialteral ratio, 3/2, rather than that of 9/4 Since it is impossible to count the vibrations of a sounding string on account of its high frequency, I should still have been in doubt as to whether a string, emitting the upper octave, made twice as many vibrations in the same time as one giving the fundamental, had it not been for the following fact, namely, that at the instant when the tone jumps to the octave, the waves which constantly accompany the vibrating glass divide up into smaller ones which are precisely half as long as the former.

Salv. This is a beautiful experiment enabling us to distinguish individually the waves which are produced by the vibrations of a sonorous body, which spread through the air, bringing to the tympanum of the ear a stimulus which the mind translates into sound. But since these waves in the water last only so long as the friction of the finger continues and are, even then, not constant but are always forming and disappearing, would it not be a fine thing if one had the ability to produce waves which would persist for a long while, even months and years, so as to easily measure and count them?

Sagr. Such an invention would, I assure you, command my admiration.

Salv. The device is one which I hit upon by accident; my part consists merely in the observation of it and in the appreciation of its value as a confirmation of something to which I had given profound consideration; and yet the device is, in itself, rather common. As I was scraping a brass plate with a sharp iron chisel in order to remove some spots from it and was running the chisel rather rapidly over it, I once or twice, during many strokes, heard the plate emit a rather strong and clear whistling sound; on looking at the plate more carefully, I noticed a long row of fine streaks parallel and equidistant from one another. Scraping with the chisel over and over again, I noticed that it was only when the plate emitted this hissing noise that any marks were left upon it; when the scraping was not accompanied by this sibilant note there was not the least trace of such marks. Repeating the trick several times and making the stroke, now with greater now with less speed, the whistling followed with a pitch which was correspondingly higher and lower. I noted also that the marks made when the tones were higher were closer together; but when the tones were deeper, they were farther apart. I also observed that when, during a single stroke, the speed increased toward the end the sound became sharper and the streaks grew closer together, but always in such a way as to remain sharply defined and equidistant. Besides whenever the stroke was accompanied by hissing I felt the chisel tremble in my grasp and a sort of shiver run through my band. In short we see and hear in the case of the chisel precisely that which is, seen and heard in the case of a whisper followed by a loud voice; for, when the breath is emitted without the production of a tone, one does not feel either in the throat or mouth any motion to speak of in comparison with that which is felt in the larynx and upper part of the throat when the voice is used, especially, when the tones employed are low and strong.

At times I have also observed among the strings of the spinet two which were in unison with two of the tones produced by the aforesaid scraping; and among those which differed most in pitch I found two which were separated by an interval of a perfect fifth. Upon measuring the distance between the markings produced by the two scrapings it was found that the space which contained 45 of one contained 30 of the other, which is precisely the ratio assigned to the fifth.

But now before proceeding any farther I want to call your attention to the fact that, of the three methods for sharpening a tone, the one which you refer to as the fineness of the string should be attributed to its weight. So long as the material of the string is unchanged, the size and weight vary in the same ratio. Thus in the case of gut-strings, we obtain the octave by making one string 4 times as large as the other; so also in the case of brass one wire must have 4 times the size of the other; but if now we wish to obtain the octave of a gut-string, by use of brass wire, we must make it, not four times as large, but four times as heavy as the gut string: as regards size therefore the metal string is not four times as big but four times as heavy. The wire may therefore be even thinner than the gut notwithstanding the fact that the latter gives the higher note. Hence if two spinets are strung, one with gold wire the other with brass, and if the corresponding strings each have the same length, diameter, and tension it follows that the instrument strung with gold will have a pitch about one-fifth lower than the other because gold has a density almost twice that of brass. And here it is to be noted that it is the weight rather than the size of a moving body which offers resistance to change of motion [*velocità del moto*] contrary to what one might at first glance think.

For it seems reasonable to believe that a body which is large and light should suffer greater retardation of motion in thrusting aside the medium than would one which is thin and heavy; yet here exactly the opposite is true.

Returning now to the original subject of discussion, I assert that the ratio of a musical interval is not immediately determined either by the length, size, or tension of the strings but rather by the ratio of their frequencies, that is, by the number of pulses of air waves which strike the tympanum of the ear, causing it also to vibrate with the same frequency. This fact established, we may possibly explain why certain pairs of notes, differing in pitch produce a pleasing sensation, others a less pleasant effect and still others a disagreeable sensation. Such an explanation would be tantamount to an explanation of the more or less perfect consonances and of dissonances. The unpleasant sensation produced by the latter arises, I think, from the discordant vibrations of two different tones which strike the ear out of time [*sproporzionatamente*]. Especially harsh is the dissonance between notes whose frequencies are incommensurable; such a case occurs when one has two strings in unison and sounds one of them open, together with a part of the other which bears the same ratio to its whole length as the side of a square bears to the diagonal; this yields a dissonance similar to the augmented fourth or diminished fifth [*tritono o semidiapente*].

Agreeable consonances are pairs of tones which strike the car with a certain regularity; this regularity consists in the fact that the pulses delivered by the two tones, in the same interval of time, shall be commensurable in number, so as not to keep the ear drum in perpetual torment, bending in two different directions in order to yield to the ever-discordant impulses.

The first and most pleasing consonance is, therefore, the octave since, for every pulse given to the tympanum by the lower string, the sharp string delivers two; accordingly at every other vibration of the upper string both pulses are delivered simultaneously so that one-half the entire number of pulses are delivered in unison. But when two strings are in unison their vibrations always coincide and the effect is that of a single string; hence we do not refer to it as consonance. The fifth is also a pleasing interval since for every two vibrations of the lower string the upper one gives three, so that considering the entire number of pulses from the upper string one-third of them will strike in unison, *i.e.*, between each pair of concordant vibrations there intervene two single vibrations; and when the interval is a fourth, three single vibrations intervene. In case the interval is a second where the ratio is 9/8 it is only every ninth vibration of the upper string which reaches the ear simultaneously with one of the lower; all the others are discordant and produce a harsh effect upon the recipient ear which interprets them as dissonances.

END OF THE FIRST DAY

THIRD DAY

CHANGE OF POSITION [DE MOTU LOCALI]

My purpose is to set forth a very new science dealing with a very ancient subject. There is, in nature, perhaps nothing older than motion, concerning which the books written by philosophers are neither few nor small; nevertheless I have discovered by experiment some properties of it which are worth knowing and which have not hitherto been either observed or demonstrated. Some superficial observations have been made, as, for instance, that the free motion [*naturalem motum*] of a heavy falling body is continuously accelerated;[1] but to just what extent this acceleration occurs has not yet been announced; for so far as I know, no one has yet pointed out that the distances traversed, during equal intervals of time, by a body falling from rest, stand to one another in the same ratio as the odd numbers beginning with unity.

It has been observed that missiles and projectiles describe a curved path of some sort; however no one has pointed out the fact that this path is a parabola. But this and other facts, not few in number or less worth knowing, I have succeeded in proving; and what I consider more important, there have been opened up to this vast and most excellent science, of which my work is merely the beginning, ways

and means by which other minds more acute than mine will explore its remote corners.

This discussion is divided into three parts; the first part deals with motion which is steady or uniform; the second treats of motion as we find it accelerated in nature; the third deals with the so-called violent motions and with projectiles.

OPPOSITE

A spacecraft's external tank falling back toward Earth illustrates the principle of naturally accelerated motion.

UNIFORM MOTION

In dealing with steady or uniform motion, we need a single definition which I give as follows:

DEFINITION

By steady or uniform motion, I mean one in which the distances traversed by the moving particle during any equal intervals of time, are themselves equal.

CAUTION

We must add to the old definition (which defined steady motion simply as one in which equal distances are traversed in equal times) the word "any," meaning by this, all equal intervals of time; for it may happen that the moving body will traverse equal distances during some equal intervals of time and yet the distances traversed during some small portion of these time-intervals may not be equal, even though the time-intervals be equal.

From the above definition, four axioms follow, namely:

AXIOM I

In the case of one and the same uniform motion, the distance traversed during a longer interval of time is greater than the distance traversed during a shorter interval of time.

AXIOM II

In the case of one and the same uniform motion, the time required to traverse a greater distance is longer than the time required for a less distance.

AXIOM III

In one and the same interval of time, the distance traversed at a greater speed is larger than the distance traversed at a less speed.

AXIOM IV

The speed required to traverse a longer distance is greater than that required to traverse a shorter distance during the same time-interval.

NATURALLY ACCELERATED MOTION

And first of all it seems desirable to find and explain a definition best fitting natural phenomena. For anyone may invent an arbitrary type of motion and discuss its properties; thus, for instance, some have imagined helices and conchoids as described by certain motions which are not met with in nature, and have very commendably established the properties which these curves possess in virtue of their definitions; but we have decided to consider the phenomena of bodies falling with an acceleration such as actually occurs in nature and to make this definition of accelerated motion exhibit the essential features of observed accelerated motions. And this, at last, after repeated efforts we trust we have succeeded in doing. In this belief we are confirmed mainly by the consideration that experimental results are seen to agree with and exactly correspond with those properties which have been, one after another, demonstrated by us. Finally, in the investigation of naturally accelerated motion we were led, by hand as it were, in following the habit and custom of nature herself, in all her various other processes, to employ only those means which are most common, simple and easy.

For I think no one believes that swimming or flying can be accomplished in a manner simpler or easier than that instinctively employed by fishes and birds.

When, therefore, I observe a stone initially at rest falling from an elevated position and continually acquiring new increments of speed, why should I not believe that such increases take place in a manner which is exceedingly simple and rather obvious to everybody? If now we examine the matter carefully we find no addition or increment more simple than that which repeats itself always in the same manner. This we readily understand when we consider the intimate relationship between time and motion; for just as uniformity of motion is defined by and conceived through equal times and equal spaces (thus we call a motion uniform when equal distances are traversed during equal time-intervals), so also we may, in a similar manner, through equal time intervals, conceive additions of speed as taking place without complication; thus we may picture to our mind a motion as uniformly and continuously accelerated when, during any equal intervals of time whatever, equal increments of speed are given to it.

Thus if any equal intervals of time whatever have elapsed, counting from the time at which the moving body left its position of rest and began to descend, the amount of speed acquired during the first two time-intervals will be double that acquired during the first time-interval alone; so the amount added during three of these time-intervals will be treble; and that in four, quadruple that of the first time-interval. To put the matter more clearly, if a body were to continue its motion with the same speed which it had acquired during the first time-interval and were to retain this same uniform speed, then its motion would be twice as slow as that which it would have if its velocity had been acquired during *two* time-intervals.

And thus, it seems, we shall not be far wrong if we put the increment of speed as proportional to the increment of time; hence the definition of motion which we are about to discuss may be stated as follows: A motion is said to be uniformly accelerated, when starting from rest, it acquires, during equal time-intervals, equal increments of speed.

Sagr. Although I can offer no rational objection to this or indeed to any other definition, devised by any author whomsoever, since all definitions are arbitrary, I may nevertheless without offense be allowed to doubt whether such a definition as the above, established in an abstract manner, corresponds to and describes that kind of accelerated motion which we meet in nature in the case of freely falling bodies. And since

the Author apparently maintains that the motion described in his definition is that of freely falling bodies, I would like to clear my mind of certain difficulties in order that I may later apply myself more earnestly to
the propositions and their demonstrations.

    Salv. It is well that you and Simplicio raise these difficulties. They are,
I imagine, the same which occurred to me when I first saw this treatise,
and which were removed either by discussion with the Author himself,
or by turning the matter over in my own mind.

    Sagr. When I think of a heavy body falling from rest, that is, starting
with zero speed and gaining speed in proportion to the time from the
beginning of the motion; such a motion as would, for instance, in eight
beats of the pulse acquire eight degrees of speed; having at the end of the

fourth beat acquired four degrees; at the end of the second, two; at the end of the first, one: and since time is divisible without limit, it follows from all these considerations that if the earlier speed of a body is less than its present speed in a constant ratio, then there is no degree of speed however small (or, one may say, no degree of slowness however great) with which we may not find this body traveling after starting from infinite slowness, i.e., from rest. So that if that speed which it had at the end of the fourth beat was such that, if kept uniform, the body would traverse two miles in an hour, and if keeping the speed which it had at the end of the second beat, it would traverse one mile an hour, we must infer that, as the instant of starting is more and more nearly approached, the body moves so slowly that, if it kept on moving at this rate, it would not traverse a mile in an hour, or in a day, or in a year or in a thousand years; indeed, it would not traverse a span in an even greater time; a phenomenon which baffles the imagination, while our senses show us that a heavy falling body suddenly acquires great speed.

Salv. This is one of the difficulties which I also at the beginning, experienced, but which I shortly afterwards removed; and the removal was effected by the very experiment which creates the difficulty for you. You say the experiment appears to show that immediately after a heavy body starts from rest it acquires a very considerable speed: And I say that the same experiment makes clear the fact that the initial motions of a falling body, no matter how heavy, are very slow and gentle. Place a heavy body upon a yielding material, and leave it there without any pressure except that owing to its own weight; it is clear that if one lifts this body a cubit or two and allows it to fall upon the same material, it will, with this impulse, exert a new and greater pressure than that caused by its mere weight; and this effect is brought about by the [weight of the] falling body together with the velocity acquired during the fall, an effect which will be greater and greater according to the height of the fall, that is according as the velocity of the falling body becomes greater. From the quality and intensity of the blow we are thus enabled to accurately estimate the speed of a falling body.

But tell me, gentlemen, is it not true that if a block be allowed to fall

89

upon a stake from a height of four cubits and drives it into the Earth, say, four finger-breadths, that coming from a height of two cubits it will drive the stake a much less distance, and from the height of one cubit a still less distance; and finally if the block be lifted only one finger-breadth how much more will it accomplish than if merely laid on top of the stake without percussion? Certainly very little. If it be lifted only the thickness of a leaf, the effect will be altogether imperceptible. And since the effect of the blow depends upon the velocity of this striking body, can any one doubt the motion is very slow and the speed more than small whenever the effect [of the blow] is imperceptible? See now the power of truth; the same experiment which at first glance seemed to show one thing, when more carefully examined, assures us of the contrary.

But without depending upon the above experiment, which is doubt-less very conclusive, it seems to me that it ought not to be difficult to establish such a fact by reasoning alone. Imagine a heavy stone held in the air at rest; the support is removed and the stone set free; then since it is heavier than the air it begins to fall, and not with uniform motion but slowly at the beginning and with a continuously accelerated motion. Now since velocity can be increased and diminished without limit, what reason is there to believe that such a moving body starting with infinite slowness, that is, from rest, immediately acquires a speed of ten degrees rather than one of four, or of two, or of one, or of a half, or of a hun-dredth; or, indeed, of any of the infinite number of small values [of speed]? Pray listen. I hardly think you will refuse to grant that the gain of speed of the stone falling from rest follows the same sequence as the diminution and loss of this same speed when, by some impelling force, the stone is thrown to its former elevation: but even if you do not grant this, I do not see how you can doubt that the ascending stone, diminish-ing in speed, must before coming to rest pass through every possible degree of slowness.

Simp. But if the number of degrees of greater and greater slowness is limitless, they will never be all exhausted, therefore such an ascending heavy body will never reach rest, but will continue to move without limit always at a slower rate; but this is not the observed fact.

Salv. This would happen, Simplicio, if the moving body were to maintain its speed for any length of time at each degree of velocity; but it merely passes each point without delaying more than an instant: and since each time-interval however small may be divided into an infinite number of instants, these will always be sufficient [in number] to correspond to the infinite degrees of diminished velocity.

That such a heavy rising body does not remain for any length of time at any given degree of velocity is evident from the following: because if, some time-interval having been assigned, the body moves with the same speed in the last as in the first instant of that time-interval, it could from this second degree of elevation be in like manner raised through an equal height, just as it was transferred from the first elevation to the second, and by the same reasoning would pass from the second to the third and would finally continue in uniform motion forever.

Sagr. From these considerations it appears to me that we may obtain a proper solution of the problem discussed by philosophers, namely, what causes the acceleration in the natural motion of heavy bodies? Since, as it seems to me, the force [virtù] impressed by the agent projecting the body upwards diminishes continuously, this force, so long as it was greater than the contrary force of gravitation, impelled the body upwards; when the two are in equilibrium the body ceases to rise and passes through the state of rest in which the impressed impetus [impeto] is not destroyed, but only its excess over the weight of the body has been consumed—the excess which caused the body to rise. Then as the diminution of the outside impetus [impeto] continues, and gravitation gains the upper hand, the fall begins, but slowly at first on account of the opposing impetus [virtù impressa], a large portion of which still remains in the body; but as this continues to diminish it also continues to be more and more overcome by gravity, hence the continuous acceleration of motion.

Simp. The idea is clever, yet more subtle than sound; for even if the argument were conclusive, it would explain only the case in which a natural motion is preceded by a violent motion, in which there still remains active a portion of the external force [virtù esterna]; but where there is no

A drawing of phases of the moon by Galileo. Galileo not only observed but also carefully recorded what he saw.

such remaining portion and the body starts from an antecedent state of rest, the cogency of the whole argument fails.

Sagr. I believe that you are mistaken and that this distinction between cases which you make is superfluous or rather non-existent. But, tell me, cannot a projectile receive from the projector either a large or a small force [*virtù*] such as will throw it to a height of a hundred cubits, and even twenty or four or one?

Simp. Undoubtedly, yes.

Sagr. So therefore this impressed force [*virtù impressa*] may exceed the resistance of gravity so slightly as to raise it only a finger-breadth; and finally the force [*virtù*] of the projector may be just large enough to exactly balance the resistance of gravity so that the body is not lifted at all but merely sustained. When one holds a stone in his hand does he do anything but give it a force impelling [*virtù impellente*] it upwards equal to the power [*facoltà*] of gravity drawing it downwards? And do you not continuously impress this force [*virtù*] upon the stone as long as you hold it in the hand? Does it perhaps diminish with the time during which one holds the stone?

And what does it matter whether this support which prevents the stone from falling is furnished by one's hand or by a table or by a rope from which it hangs? Certainly nothing at all. You must conclude, therefore, Simplicio, that it makes no difference whatever whether the fall of the stone is preceded by a period of rest which is long, short, or instantaneous provided only the fall does not take place so long as the stone is acted upon by a force [*virtù*] opposed to its weight and sufficient to hold it at rest.

Salv. The present does not seem to be the proper time to investigate the cause of the acceleration of natural motion concerning which various opinions have been expressed by various philosophers, some explaining it by attraction to the center, others to repulsion between the very small parts of the body, while still others attribute it to a certain stress in the surrounding medium which closes in behind the falling body and drives it from one of its positions to another. Now, all these fantasies, and others too, ought to be examined; but it is not really worth while. At

present it is the purpose of our Author merely to investigate and to demonstrate some of the properties of accelerated motion (whatever the cause of this acceleration may be)—meaning thereby a motion, such that the momentum of its velocity [*i momenti della sua velocità*] goes on increasing after departure from rest, in simple proportionality to the time, which is the same as saying that in equal time-intervals the body receives equal increments of velocity; and if we find the properties [of accelerated motion] which will be demonstrated later are realized in freely falling and accelerated bodies, we may conclude that the assumed definition includes such a motion of falling bodies and that their speed [*accelerazione*] goes on increasing as the time and the duration of the motion.

Sagr. So far as I see at present, the definition might have been put a little more clearly perhaps without changing the fundamental idea, namely, uniformly accelerated motion is such that its speed increases in proportion to the space traversed; so that, for example, the speed acquired by a body in falling four cubits would be double that acquired in falling two cubits and this latter speed would be double that acquired in the first cubit. Because there is no doubt but that a heavy body falling from the height of six cubits has, and strikes with, a momentum [*impeto*] double that it had at the end of three cubits, triple that which it had at the end of one.

Salv. It is very comforting to me to have had such a companion in error; and moreover let me tell you that your proposition seems so highly probable that our Author himself admitted, when I advanced this opinion to him, that he had for some time shared the same fallacy. But what most surprised me was to see two propositions so inherently probable that they commanded the assent of everyone to whom they were presented, proven in a few simple words to be not only false, but impossible.

Simp. I am one of those who accept the proposition, and believe that a falling body acquires force [*vires*] in its descent, its velocity increasing in proportion to the space, and that the momentum [*momento*] of the falling body is doubled when it falls rom a doubled height; these propositions, it appears to me, ought to be conceded without hesitation or controversy.

Salv. And yet they are as false and impossible as that motion should be completed instantaneously; and here is a very clear demonstration of it. If the velocities are in proportion to the spaces traversed, or to be traversed, then these spaces are traversed in equal intervals of time; if, therefore, the velocity with which the falling body traverses a space of eight feet were double that with which it covered the first four feet (just as the one distance is double the other) then the time-intervals required for these passages would be equal. But for one and the same body to fall eight feet and four feet in the same time is possible only in the case of instantaneous [discontinuous] motion; but observation shows us that the motion of a falling body occupies time, and less of it in covering a distance of four feet than of eight feet; therefore it is not true that its velocity increases in proportion to the space.

The falsity of the other proposition may be shown with equal clearness. For if we consider a single striking body the difference of momentum in its blows can depend only upon difference of velocity; for if the striking body falling from a double height were to deliver a blow of double momentum, it would be necessary for this body to strike with a doubled velocity; but with this doubled speed it would traverse a doubled space in the same time-interval; observation however shows that the time required for fall from the greater height is longer.

Sagr. You present these recondite matters with too much evidence and ease; this great facility makes them less appreciated than they would be had they been presented in a more abstruse manner. For, in my opinion, people esteem more lightly that knowledge which they acquire with so little labor than that acquired through long and obscure discussion.

Salv. If those who demonstrate with brevity and clearness the fallacy of many popular beliefs were treated with contempt instead of gratitude the injury would be quite bearable; but on the other hand it is very unpleasant and annoying to see men, who claim to be peers of anyone in a certain field of study, take for granted certain conclusions which later arc quickly and easily shown by another to be false. I do not describe such a feeling as one of envy, which usually degenerates into hatred and anger against those who discover such fallacies; I would call it a strong desire to maintain old errors, rather than accept newly discovered truths.

OPPOSITE PAGE

The Hubble space telescope.
This is the twenty-first century
version of Galileo's telescope and
continues the nature of observation
that exemplifies the theoretical
models created in his time
and thereafter.

This desire at times induces them to unite against these truths, although at heart believing in them, merely for the purpose of lowering the esteem in which certain others are held by the unthinking crowd. Indeed, I have heard from our Academician many such fallacies held as true but easily refutable; some of these I have in mind.

Sagr. You must not withhold them from us, but, at the proper time, tell us about them even though an extra session be necessary. But now, continuing the thread of our talk, it would seem that up to the present we have established the definition of uniformly accelerated motion which is expressed as follows:

A motion is said to be equally or uniformly accelerated when, starting from rest, its momentum (celeritatis momenta) receives equal increments in equal times.

Salv. This definition established, the Author makes a single assumption, namely,

The speeds acquired by one and the same body moving down planes of different inclinations are equal when the heights of these planes are equal.

————

END OF THE THIRD DAY

————

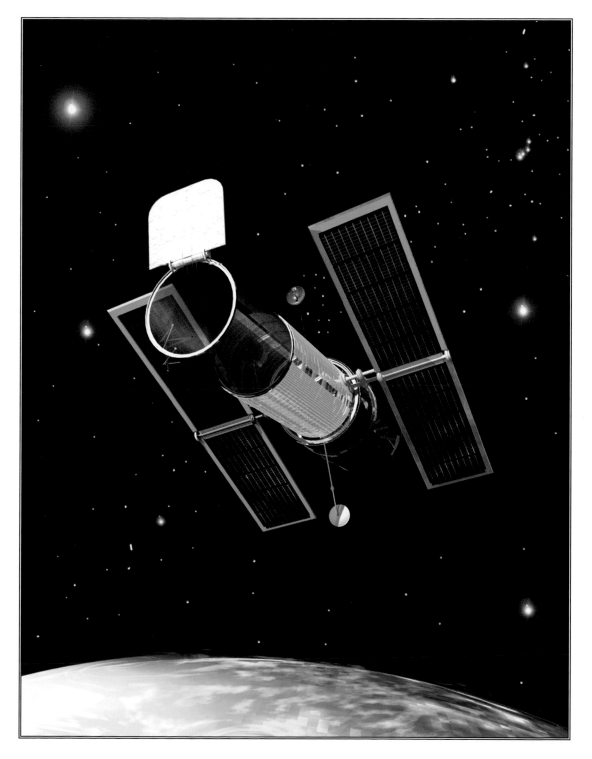

Johannes Kepler (1571-1630)

HIS LIFE AND WORK

If an award were ever given to the person in history who was most dedicated to the pursuit of absolute precision, the German astronomer Johannes Kepler might well be the recipient. Kepler was so obsessed with measurements that he even calculated his own gestational period to the minute—224 days, 9 hours, 53 minutes. (He had been born prematurely.) So it is no surprise that he toiled over his astronomical research to such a degree that he ultimately produced the most exact astronomical tables of his time, leading to the eventual acceptance of the Sun-centered (heliocentric) theory of the planetary system.

Like Copernicus, whose work inspired him, Kepler was a deeply religious man. He viewed his continual study of universal properties as a fulfillment of his Christian duty to understand the very universe that God created. But unlike Copernicus, Kepler's life was anything but quiet and lacking in contrast. Always short of money, Kepler often resorted to publishing astrological calendars and horoscopes, which, ironically, gained him some local notoriety when their predictions turned out to be quite accurate. Kepler also suffered the early deaths of several of his children, as well as the indignity of having to defend in court his eccentric mother, Katherine, who had a reputation for practicing witchcraft and was nearly burned at the stake.

Kepler entered into a series of complex relationships, most notably with Tycho Brahe, the great naked-eye astronomical observer. Brahe dedicated years of his life to recording and measuring celestial bodies, but he lacked the mathematical and analytical skills necessary to understand

Danish astronomer Tycho Brahe, Kepler's employer.

planetary motion. A man of wealth, Brahe hired Kepler to make sense of his observations of the orbit of Mars, which had perplexed astronomers for many years. Kepler painstakingly mapped Brahe's data on the motion of Mars to an ellipse, and this success lent mathematical credibility to the Copernican model of a Sun-centered system. His discovery of elliptical orbits helped usher in a new era in astronomy. The motions of planets could now be predicted.

In spite of his achievements, Kepler never gained much wealth or prestige and was often forced to flee the countries where he sojourned because of religious upheaval and civil unrest. By the time he died at the age of fifty-nine in 1630 (while attempting to collect an overdue salary), Kepler had discovered three laws of planetary motion, which are still taught to students in physics classes in the twenty-first century. And it was Kepler's Third Law, not an apple, that led Isaac Newton to discover the law of gravitation.

Johannes Kepler was born on December 27, 1571, in the town of Weil der Stadt, in Württemburg (now part of Germany). His father, Heinrich Kepler, was, according to Johannes, "an immoral, rough, and quarrelsome soldier" who deserted his family on several occasions to join up with mercenaries to battle a Protestant uprising in Holland. Heinrich is believed to have died somewhere in the Netherlands. The young Johannes lived with his mother, Katherine, in his grandfather's inn, where he was put to work at an early age waiting tables, despite his poor health. Kepler had nearsightedness as well as double vision, which was believed to have been caused by a near-fatal bout of smallpox; and he also suffered from abdominal problems and "crippled" fingers that limited his career potential choice, in the view of his family, to a life in the ministry.

"Bad-tempered" and "garrulous" were words Kepler used to describe his mother, Katherine, but he was aware from a young age that his father was the cause. Katherine herself had been raised by an aunt who practiced witchcraft and was burned at the stake. So it was no surprise to Kepler when his own mother faced similar charges later in her life. In 1577, Katherine showed her son the "great comet" that appeared in the sky that year, and Kepler later acknowledged that this shared moment

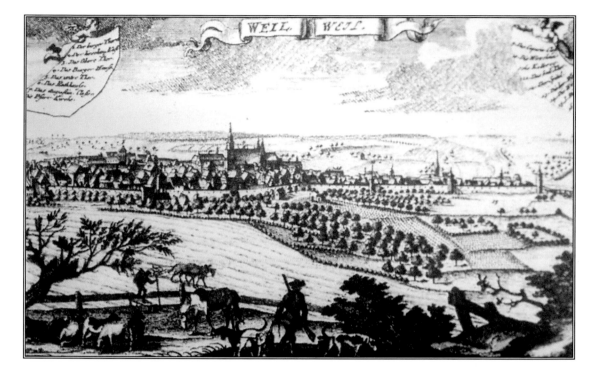

WEIL. WEIL.

with his mother had a lasting impact on his life. Despite a childhood filled with pain and anxiety, Kepler was obviously gifted, and he managed to procure a scholarship reserved for promising male children of limited means who lived in the German province of Swabia. He attended the German Schreibschule in Leonberg before transferring to a Latin school, which was instrumental in providing him with the Latin writing style he later employed in his work. Being frail and precocious, Kepler was beaten regularly by classmates, who considered him a know-it-all, and he soon turned to religious study as a way of escaping his predicament.

In 1587, Kepler enrolled at Tübingen University, where he studied theology and philosophy. He also established himself there as a serious student of mathematics and astronomy, and became an advocate of the controversial Copernican heliocentric theory. So public was young Kepler in his defense of the Copernican model of the universe that it was not uncommon for him to engage in public debate on the subject. Despite his main interest in theology, he was growing more and more intrigued by the mystical appeal of a heliocentric universe. Although he

The Imperial city of Weil der Stadt, Germany, where Kepler was born.

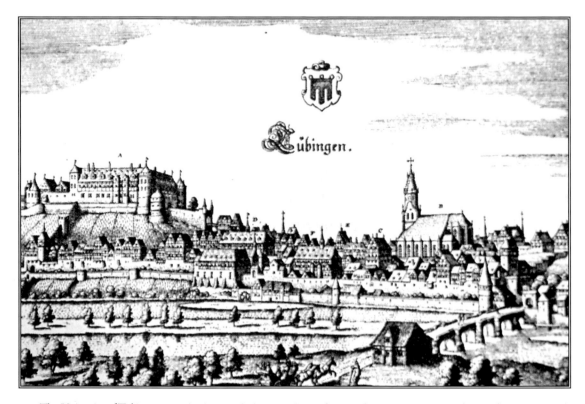

The University of Tübingen. Kepler studied here for a master's degree in theology.

had intended to graduate from Tubingen in 1591 and join the university's theology faculty, a recommendation to a post in mathematics and astronomy at the Protestant school in Graz, Austria, proved irresistible. So, at the age of twenty-two, Kepler deserted a career in the ministry for the study of science. But he would never abandon his belief in God's role in the creation of the universe.

In the sixteenth century, the distinction between astronomy and astrology was fairly ambiguous. One of Kepler's duties as a mathematician in Graz was to compose an astrological calendar complete with predictions. This was a common practice at the time, and Kepler was clearly motivated by the extra money the job provided, but he could not have anticipated the public's reaction when his first calendar was published. He predicted an extraordinarily cold winter, as well as a Turkish incursion, and when both predictions came true, Kepler was triumphantly hailed as a prophet. Despite the clamor, he would never hold much respect for the work he did on the annual almanacs. He called

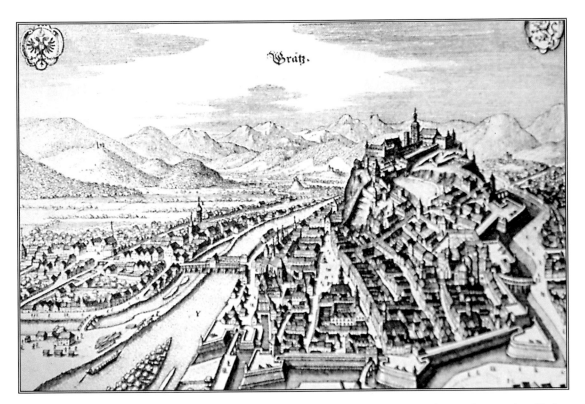

The city of Graz, where Kepler became a teacher in a seminary after completing his studies.

astrology "the foolish little daughter of astronomy" and was equally dismissive of the public's interest and the astrologer's intentions. "If ever astrologers are correct," he wrote, "it ought to be credited to luck." Still, Kepler never failed to turn to astrology whenever money became tight, which was a recurring theme in his life, and he did hold out hope of discovering some true science in astrology.

One day, while lecturing on geometry in Graz, Kepler experienced a sudden revelation that set him on a passionate journey and changed the course of his life. It was, he felt, the secret key to understanding the universe. On the blackboard, in front of the class, he drew an equilateral triangle within a circle, and another circle drawn within the triangle. It occurred to him that the ratio of the circles was indicative of the ratio of the orbits of Saturn and Jupiter. Inspired by this revelation, he assumed that all six planets known at the time were arranged around the Sun in such a way that the geometric figures would fit perfectly between them. Initially he tested this hypothesis without success, using two-dimensional

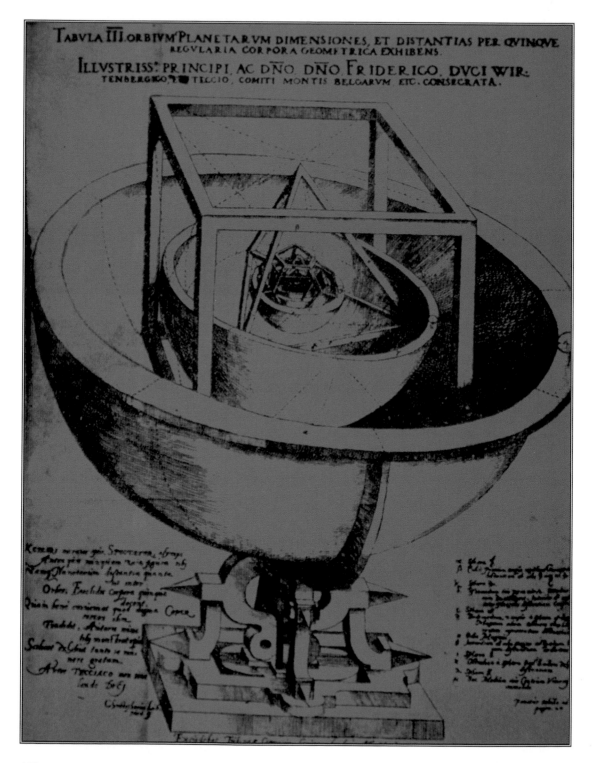

plane figures such as the pentagon, the square, and the triangle. He then returned to the Pythagorean solids, used by the ancient Greeks, who discovered that only five solids could be constructed from regular geometric figures. To Kepler, this explained why there could only be six planets (Mercury, Venus, Earth, Mars, Jupiter, and Saturn) with five spaces between them, and why these spaces were not uniform. This geometric theory regarding planetary orbits and distances inspired Kepler to write *Mystery of the Cosmos* (*Mysterium Cosmographicum*), published in 1596. It took him about a year to write, and although the scheme was reasonably accurate, he was clearly very sure that his theories would ultimately bear out:

And how intense was my pleasure from this discovery can never be expressed in words. I no longer regretted the time wasted. Day and night I was consumed by the computing, to see whether this idea would agree with the Copernican orbits, or if my joy would be carried away by the wind. Within a few days everything worked, and I watched as one body after another fit precisely into its place among the planets.

Kepler spent the rest of his life trying to obtain the mathematical proof and scientific observations that would justify his theories. *Mystery of the Cosmos* was the first decidedly Copernican work published since Copernicus' own *On the Revolutions*, and as a theologian and astronomer Kepler was determined to understand how and why God designed the universe. Advocating a heliocentric system had serious religious implications, but Kepler maintained that the sun's centrality was vital to God's design, as it kept the planets aligned and in motion. In this sense, Kepler broke with Copernicus' heliostatic system of a Sun "near" the center and placed the Sun directly in the center of the system.

Today, Kepler's polyhedra appear impracticable. But although the premise of *Mystery of the Cosmos* was erroneous, Kepler's conclusions were still astonishingly accurate and decisive, and were essential in shaping the course of modern science. When the book was published, Kepler sent a copy to Galileo, urging him to "believe and step forth," but the

OPPOSITE PAGE

Kepler's drawing of his model of the five platonic solids.

Italian astronomer rejected the work because of its apparent speculations. Tycho Brahe, on the other hand, was immediately intrigued. He viewed Kepler's work as new and exciting, and he wrote a detailed critique in the book's support. Reaction to *Mystery of the Cosmos*, Kepler would later write, changed the direction of his entire life.

In 1597, another event would change Kepler's life, as he fell in love with Barbara Müller, the first daughter of a wealthy mill owner. They married on April 27 of that year, under an unfavorable constellation, as Kepler would later note in his diary. Once again, his prophetic nature emerged as the relationship and the marriage dissolved. Their first two children died very young, and Kepler became distraught. He immersed himself in his work to distract himself from the pain, but his wife did not understand his pursuits. "Fat, confused, and simpleminded" was how he described her in his diary, though the marriage did last fourteen years, until her death in 1611 from typhus.

Kepler's first wife, Barbara. They were married in 1597.

In September 1598, Kepler and other Lutherans in Graz were ordered to leave town by the Catholic archduke, who was bent on removing the Lutheran religion from Austria. After a visit to Tycho Brahe's Benatky Castle in Prague, Kepler was invited by the wealthy Danish astronomer to stay there and work on his research. Kepler was somewhat wary of Brahe, even before having met him. "My opinion of Tycho is this: he is superlatively rich, but he knows not how to make proper use of it, as is the case with most rich people," he wrote. "Therefore, one must try to wrest his riches from him."

If his relationship with his wife lacked complexity, Kepler more than made up for it when he entered into a working arrangement with the aristocratic Brahe. At first, Brahe treated the young Kepler as an assistant, carefully doling out assignments without giving him much access to detailed observational data. Kepler badly wanted to be regarded as an equal and given some independence, but the secretive Brahe wanted to use Kepler to establish his own model of the solar system—a non-Copernican model that Kepler did not support.

The young Kepler.

Kepler was immensely frustrated. Brahe had a wealth of observational data but lacked the mathematical tools to fully comprehend it. Finally, perhaps to pacify his restless assistant, Brahe assigned Kepler to study the orbit of Mars, which had confused the Danish astronomer for some time, because it appeared to be the least circular. Kepler initially thought he could solve the problem in eight days, but the project turned out to take him eight years. Difficult as the research proved to be, it was not without its rewards, as the work led Kepler to discover that Mars's orbit precisely described an ellipse, as well as to formulate his first two "planetary laws," which he published in 1609 in *The New Astronomy*.

A year and a half into his working relationship with Brahe, the Danish astronomer became very ill at dinner and died a few days later of a bladder infection. Kepler took over the post of Imperial Mathematician and was now free to explore planetary theory without being constrained by the watchful eye of Tycho Brahe. Realizing an opportunity, Kepler immediately went after the Brahe data that he coveted before Brahe's

Kepler and Brahe from an eighteenth-century German atlas.

heirs could take control of them. "I confess that when Tycho died," Kepler wrote later, "I quickly took advantage of the absence, or lack of circumspection, of the heirs, by taking the observations under my care, or perhaps usurping them." The result was Kepler's *Rudolphine Tables*, a compilation of the data from thirty years of Brahe's observations. To be fair, on his deathbed Brahe had urged Kepler to complete the tables; but Kepler did not frame the work according to any Tychonic hypothesis, as Brahe had hoped. Instead, Kepler used the data, which included calculations using logarithms he had developed himself, in predicting planetary positions. He was able to predict transits of the sun by Mercury and Venus, though he did not live long enough to witness them. Kepler did not publish *Rudolphine Tables* until 1627, however, because the data he discovered constantly led him in new directions.

After Brahe's death, Kepler witnessed a nova, which later became known as "Kepler's nova," and he also experimented in optical theories. Though scientists and scholars view Kepler's optical work as minor in comparison with his accomplishments in astronomy and mathematics, the publication in 1611 of his book *Dioptrices*, changed the course of optics.

In 1605, Kepler announced his first law, the law of ellipses, which held that the planets move in ellipses with the Sun at one focus. Earth, Kepler asserted, is closest to the Sun in January and farthest from it in July as it travels along its elliptical orbit. His second law, the law of equal areas, maintained that a line drawn from the Sun to a planet sweeps out equal areas in equal times. Kepler demonstrated this by arguing that an imaginary line connecting any planet to the Sun must sweep over equal areas in equal intervals of time. He published both laws in 1609 in his book *New Astronomy (Astronomia Nova)*.

Yet despite his status as Imperial Mathematician and as a distinguished scientist whom Galileo sought out for an opinion on his new telescopic discoveries, Kepler was unable to secure for himself a comfortable existence. Religious upheaval in Prague jeopardized his new homeland, and in 1611 his wife and his favorite son died. Kepler was permitted, under exemption, to return to Linz, and in 1613 he married Susanna Reuttinger, a twenty-four-year-old orphan who would bear him seven children, only two of whom would survive to adulthood. It was at this time that Kepler's mother was accused of witchcraft, and in the midst of his own personal turmoil he was forced to defend her against the charge in order to prevent her being burned at the stake. Katherine was imprisoned and tortured, but her son managed to obtain an acquittal, and she was released.

Because of these distractions, Kepler's return to Linz was not a productive time initially. Distraught, he turned his attention away from tables and began working on *Harmonies of the World (Harmonice Mundi)*, a passionate work which Max Caspar, in his biography of Kepler, described as "a great cosmic vision, woven out of science, poetry, philosophy, theology,

mysticism." Kepler finished *Harmonies of the World* on May 27, 1618. In this series of five books, he extended his theory of harmony to music, astrology, geometry, and astronomy. The series included his third law of planetary motion, the law that would inspire Isaac Newton some sixty years later, which maintained that the cubes of mean distances of the planets from the Sun are proportional to the squares of their periods of revolution. In short, Kepler discovered how planets orbited, and in so doing paved the way for Newton to discover why.

Kepler believed he had discovered God's logic in designing the universe, and he was unable to hide his ecstasy. In Book 5 of *Harmonies of the World* he wrote:

I dare frankly to confess that I have stolen the golden vessels of the Egyptians to build a tabernacle for my God far from the bounds of Egypt. If you pardon me, I shall rejoice; if you reproach me, I shall endure. The die is cast, and I am writing the book, to be read either now or by posterity, it matters not. It can wait a century for a reader, as God himself has waited six thousand years for a witness.

The Thirty Years War, which beginning in 1618 decimated the Austrian and German lands, forced Kepler to leave Linz in 1626. He eventually settled in the town of Sagan, in Silesia. There he tried to finish what might best be described as a science fiction novel, which he had dabbled at for years, at some expense to his mother during her trial for witchcraft. *Dream of the Moon* (*Somnium seu astronomia lunari*), which features an interview with a knowing "demon" who explains how the protagonist could travel to the moon, was uncovered and presented as evidence during Katherine's trial. Kepler spent considerable energy defending the work as pure fiction and the demon as a mere literary device. The book was unique in that it was not only ahead of its time in terms of fantasy but also a treatise supporting Copernican theory.

In 1630, at the age of fifty-eight, Kepler once again found himself in financial straits. He set out for Regensburg, where he hoped to collect interest on some bonds in his possession as well as some money he was

This globe from the Uraniborg library was begun in Augburg in 1570 and completed ten years later.

owed. However, a few days after his arrival he developed a fever, and died on November 15. Though he never achieved the mass renown of Galileo, Kepler produced a body of work that was extraordinarily useful to professional astronomers like Newton who immersed themselves in the details and accuracy of Kepler's science. Johannes Kepler was a man who preferred aesthetic harmony and order, and all that he discovered was inextricably linked with his vision of God. His epitaph, which he himself composed, reads: "I used to measure the heavens; now I shall measure the shadows of the earth. Although my soul was from heaven, the shadow of my body lies here."

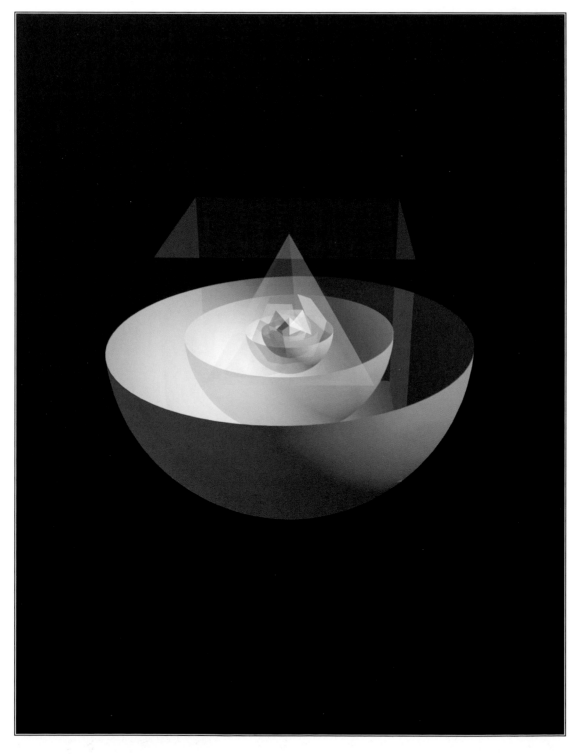

HARMONIES OF THE WORLD

BOOK FIVE

Concerning the very perfect harmony of the celestial movements, and the genesis of eccentricities and the semidiameters, and as periodic times from the same.

After the model of the most correct astronomical doctrine of today, and the hypothesis not only of Copernicus but also of Tycho Brahe, whereof either hypotheses are today publicly accepted as most true, and the Ptolemaic as outmoded.

I commence a sacred discourse, a most true hymn to God the Founder, and I judge it to be piety, not to sacrifice many hecatombs of bulls to Him and to burn incense of innumerable perfumes and cassia, but first to learn myself, and afterwards to teach others too, how great He is in wisdom, how great in power, and of what sort in goodness. For to wish to adorn in every way possible the things that should receive adornment and to envy no thing its goods—this I put down as the sign of the greatest goodness, and in this respect I praise Him as good that in the heights of His wisdom He finds everything whereby each thing may be adorned to the utmost and that He can do by His unconquerable power all that He has decreed.

Galen, *on the Use of Parts.* Book III

PROEM

As regards that which I prophesied two and twenty years ago (especially that the five regular solids are found between the celestial spheres), as regards that of which I was firmly persuaded in my own mind before I had seen Ptolemy's *Harmonies,* as regards that which I promised my friends in the title of this fifth book before I was sure of the thing itself, that which, sixteen years ago, in a published statement, I insisted must be investigated, for the sake of which I spent the best part of my life in astronomical speculations, visited Tycho Brahe, and took up residence at Prague: finally, as God the Best and Greatest, Who had inspired my mind

OPPOSITE PAGE

Harmonies within the universe. The structure of the universe is seen here as a series of nesting units comprising the five platonic solids. The sphere contains the cube, contains the sphere, contains the tetrahedron, contains the sphere contains the octahedron contains the sphere, contains the dodecahedron contains the sphere contains the icosahedron.

and aroused my great desire, prolonged my life and strength of mind and furnished the other means through the liberality of the two Emperors and the nobles of this province of Austria-on-the-Anisana: after I had discharged my astronomical duties as much as sufficed, finally, I say, I brought it to light and found it to be truer than I had even hoped, and I discovered among the celestial movements the full nature of harmony, in its due measure, together with all its parts unfolded in Book III—not in that mode wherein I had conceived it in my mind (this is not last in my joy) but in a very different mode which is also very excellent and very perfect. There took place in this intervening time, wherein the very laborious reconstruction of the movements held me in suspense, an extraordinary augmentation of my desire and incentive for the job, a reading of the *Harmonies* of Ptolemy, which had been sent to me in manuscript by John George Herward, Chancellor of Bavaria, a very distinguished man and of a nature to advance philosophy and every type of learning. There, beyond my expectations and with the greatest wonder, I found approximately the whole third book given over to the same consideration of celestial harmony, fifteen hundred years ago. But indeed astronomy was far from being of age as yet; and Ptolemy, in an unfortunate attempt, could make others subject to despair, as being one who, like Scipio in Cicero, seemed to have recited a pleasant Pythagorean dream rather than to have aided philosophy. But both the crudeness of the ancient philosophy and this exact agreement in our meditations, down to the last hair, over an interval of fifteen centuries, greatly strengthened me in getting on with the job. For what need is there of many men? The very nature of things, in order to reveal herself to mankind, was at work in the different interpreters of different ages, and was the finger of God—to use the Hebrew expression; and here, in the minds of two men, who had wholly given themselves up to the contemplation of nature, there was the same conception as to the configuration of the world, although neither had been the other's guide in taking this route. But now since the first light eight months ago, since broad day three months ago, and since the sun of my wonderful speculation has shone fully a very few days ago: nothing holds me back. I am free to give myself up to the sacred

Tycho Brahe's quadrant, used in his observatory at Uraniborg.

madness, I am free to taunt mortals with the frank confession that I am stealing the golden vessels of the Egyptians, in order to build of them a temple for my God, far from the territory of Egypt. If you pardon me, I shall rejoice; if you are enraged, I shall bear up. The die is cast, and I am writing the book—whether to be read by my contemporaries or by posterity matters not. Let it await its reader for a hundred years, if God Himself has been ready for His contemplator for six thousand years.

———

Before taking up these questions, it is my wish to impress upon my readers the very exhortation of Timaeus, a pagan philosopher, who was going to speak on the same things: it should be learned by Christians with the greatest admiration, and shame too, if they do not imitate him:

———

For truly, Socrates, since all who have the least particle of intelligence always invoke God whenever they enter upon any business, whether light or arduous; so too, unless we have clearly strayed away from all sound reason, we who intend to have a discussion concerning the universe must of necessity make our sacred wishes and pray to the Gods and Goddesses with one mind that we may say such things as will please and be acceptable to them in especial and, secondly, to you too.

Kepler's view of the universe linked the planets with the platonic solids and their cosmic geometries. Mars as dodecahedron, Venus as icosahedron, Earth as sphere, Jupiter as tetrahedron, Mercury as octahedron, Saturn as cube.

I. CONCERNING THE FIVE REGULAR SOLID FIGURES

It has been said in the second book how the regular plane figures are fitted together to form solids; there we spoke of the five regular solids, among others, on account of the plane figures. Nevertheless their number, five, was there demonstrated; and it was added why they were designated by the Platonists as the figures of the world, and to what element any solid was compared on account of what property. But now, in the anteroom of this book, I must speak again concerning these figures, on their own account, not on account of the planes, as much as suffices for the celestial harmonies; the reader will find the rest in the *Epitome of Astronomy*, Volume II, Book IV.

Accordingly, from the *Mysterium Cosmographicum*, let me here briefly inculcate the order of the five solids in the world, whereof three are primary and two secondary. For the *cube* (1) is the outmost and the most spacious, because firstborn and having the nature (*rationem*) of a *whole*, in the very form of its generation. There follows the *tetrahedron* (2), as if made a *part*, by cutting up the cube; nevertheless it is primary too, with a solid trilinear angle, like the cube. Within the tetrahedron is the *dodecahedron* (3), the last of primary figures, namely, like a solid composed of parts of a cube and similar parts of a tetrahedron, *i.e.*, of irregular tetrahedrons, wherewith the cube inside is roofed over. Next in order is the *icosahedron* (4) on account of its similarity, the last of the secondary figures and having a plurilinear solid angle. The *octahedron* (5) is inmost, which is similar to the cube and the first of the secondary figures and to which as inscriptile the first place is due, just as the first outside place is due to the cube as circumscriptile.

However, there are as it were two noteworthy weddings of these figures, made from different classes: the males, the cube and the dodecahedron, among the primary; the females, the octahedron and the icosahedron, among the secondary, to which is added one as it were bachelor or hermaphrodite, the tetrahedron, because it is inscribed in itself, just as those female solids are inscribed in the males and are as it were subject to them, and have the signs of the feminine sex, opposite the masculine, namely, angles opposite planes. Moreover, just as the tetrahedron is the element, bowels, and as it were rib of the male cube, so the feminine octahedron is the element and part of the tetrahedron in another way; and thus the tetrahedron mediates in this marriage.

The main difference in these wedlocks or family relationships consists in the following: The ratio of the cube is *rational*. For the tetrahedron is one third of the body of the cube, and the octahedron half of the tetrahedron, one sixth of the cube; while the ratio of the dodecahedron's wedding is *irrational (ineffabilis)* but *divine*.

The union of these two words commands the reader to be careful as to their significance. For the word *ineffabilis* here does not of itself denote any nobility, as elsewhere in theology and divine things, but denotes an inferior condition. For in geometry, as was said in the first book, there are many irrationals, which do not on that account participate in a divine proportion too. But you must look in the first book for what the divine ratio, or rather the divine section, is. For in other proportions there are four terms present; and three, in a continued proportion; but the divine requires a single relation of terms outside of that of the proportion itself, namely in such fashion that the two lesser terms, as parts make up the greater term, as a whole. Therefore, as much as is taken away from this wedding of the dodecahedron on account of its employing an irrational proportion, is added to it conversely, because its irrationality approaches the divine. This wedding also comprehends the solid star too, the generation whereof arises from the continuation of five planes of the dodecahedron till they all meet in a single point. See its generation in Book II.

Lastly, we must note the ratio of the spheres circumscribed around them to those inscribed in them: in the case of the tetrahedron it is rational, 100,000:33,333 or 3:1; in the wedding of the cube it is irrational, but the radius of the inscribed sphere is rational in square, and is itself the square root of one third the square on the radius (of the circumscribed sphere), namely 100,000:57,735; in the wedding of the dodecahedron, clearly irrational, 100,000:79,465; in the case of the star, 100,000:52,573, half the side of the icosahedron or half the distance between two rays.

2. ON THE KINSHIP BETWEEN THE HARMONIC RATIOS AND THE FIVE REGULAR FIGURES

This kinship (*cognatio*) is various and manifold; but there are four degrees of kinship. For either the sign of kinship is taken from the outward form alone which the figures have, or else ratios which are the same as the harmonic arise in the construction of the side, or result from the figures already constructed, taken simply or together; or, lastly, they are either equal to or approximate the ratios of the spheres of the figure.

In the first degree, the ratios, where the character or greater term is 3, have kinship with the triangular plane of the tetrahedron, octahedron, and icosahedron; but where the greater term is 4, with the square plane of the cube; where 5, with the pentagonal plane of the dodecahedron. This similitude on the part of the plane can also be extended to the smaller term of the ratio, so that wherever the number 3 is found as one

term of the continued doubles, that ratio is held to be akin to the three figures first named: for example, 1:3 and 2:3 and 4:3 and 8:3, et cetera; but where the number is 5, that ratio is absolutely assigned to the wedding of the dodecahedron: for example, 2:5 and 4:5 and 8:5, and thus 3:5 and 3:10 and 6:5 and 12:5 and 24:5. The kinship will be less probable if the sum of the terms expresses this similitude, as in 2:3 the sum of the

The five platonic solids that
Kepler believed to be the
building blocks of the Universe.
The sphere contains them all
(as shown in the reflected crystal).

terms is equal to 5, as if to say that 2:3 is akin to the dodecahedron. The
kinship on account of the outward form of the solid angle is similar: the
solid angle is trilinear among the primary figures, quadrilinear in the
octahedron, and quinquelinear in the icosahedron. And so if one term of
the ratio participates in the number 3, the ratio will be connected with
the primary bodies; but if in the number 4, with the octahedron; and

finally, if in the number 5, with the icosahedron. But in the feminine solids this kinship is more apparent, because the characteristic figure latent within follows upon the form of the angle: the tetragon in the octahedron, the pentagon in the icosahedron; and so 3:5 would go to the sectioned icosahedron for both reasons.

The second degree of kinship, which is genetic, is to be conceived as follows: First, some harmonic ratios of numbers are akin to one wedding or family, namely, perfect ratios to the single family of the cube; conversely, there is the ratio which is never fully expressed in numbers and cannot be demonstrated by numbers in any other way, except by a long series of numbers gradually approaching it: this ratio is called *divine*, when it is perfect, and it rules in various ways throughout the dodecahedral wedding. Accordingly, the following consonances begin to shadow forth that ratio: 1:2 and 2:3 and 2:3 and 5:8. For it exists most imperfectly in 1:2, more perfectly in 5:8, and still more perfectly if we add 5 and 8 to make 13 and take 8 as the numerator, if this ratio has not stopped being harmonic.

Further, in constructing the side of the figure, the diameter of the globe must be cut; and the octahedron demands its bisection, the cube and the tetrahedron its trisection, the dodecahedral wedding its quinquesection. Accordingly, the ratios between the figures are distributed according to the numbers which express those ratios. But the square on the diameter is cut too, or the square on the side of the figure is formed from a fixed part of the diameter. And then the squares on the sides are compared with the square on the diameter, and they constitute the following ratios: in the cube 1:3, in the tetrahedron 2:3, in the octahedron 1:2. Wherefore, if the two ratios are put together, the cubic and the tetrahedral will give 1:2; the cubic and the octahedral, 2:3; the octahedral and the tetrahedral, 3:4. The sides in the dodecahedral wedding are irrational.

Thirdly, the harmonic ratios follow in various ways upon the already constructed figures. For either the number of the sides of the plane is compared with the number of lines in the total figure; and the following ratios arise: in the cube 4:12 or 1:3; in the tetrahedron 3:6 or 1:2; in the octahedron 3:12 or 1:4; in the dodecahedron 5:30 or 1:6; in the icosahedron 3:30 or 1:10. Or else the number of sides of the plane is compared with the number of planes; then the cube gives 4:6 or 2:3, the tetrahedron 3:4, the octahedron 3:8, the dodecahedron 5:12, the icosahedron 3:20. Or else the number of sides or angles of the plane is compared with the number of solid angles, and the cube gives 4:8 or 1:2, the tetrahedron 3:4, the octahedron 3:6 or 1:2, the dodecahedron with its consort 5:20 or 3:12 (*i.e.*, 1:4). Or else the number of planes is compared with the number of solid angles, and the cubic wedding gives 6:8 or 3:4, the tetrahedron the ratio of equality, the dodecahedral wedding 1:20 or 3:5. Or else the number of all the sides is compared with the number of the solid angles, and the cube gives 8:12 or 2:3, the tetrahedron 4:6 or 2:3, and the octahedron 6:12 or 1:2, the dodecahedron 20:30 or 2:3, the icosahedron 12:30 or 2:5.

Moreover, the bodies too are compared with one another, if the tetrahedron is stowed away in the cube, the octahedron in the tetrahedron and cube, by geometrical inscription. The tetrahedron is one third of the cube, the octahedron half of the tetrahedron, one sixth of the cube, just as the octahedron, which is inscribed in the globe, is one sixth of the cube which circumscribes the globe. The ratios of the remaining bodies are irrational.

The fourth species or degree of kinship is more proper to this work: the ratio of the spheres inscribed in the figures to the spheres circumscribing them is sought, and what harmonic ratios approximate them is calculated. For only in the tetrahedron is the diameter of the inscribed sphere rational, namely, one third of the circumscribed sphere. But in the cubic wedding the ratio, which is single there, is as lines which are rational only in square. For the diameter of the inscribed sphere is to the diameter of the circumscribed sphere as the square root of the ratio 1:3. And if you compare the ratios with one another, the ratio of the tetrahedral spheres is the square of the ratio of the cubic spheres. In the dodecahedral wedding there is again a single ratio, but an irrational one, slightly greater than 4:5. Therefore the ratio of the spheres of the cube and octahedron is approximated by the following consonances: 1:2, as proximately greater, and 3:5, as proximately smaller. But the ratio of the dodecahedral spheres is approximated by the consonances 4:5 and 5:6, as proximately smaller, and 3:4 and 5:8, as proximately greater.

But if for certain reasons 1:2 and 1:3 are arrogated to the cube, the ratio of the spheres of the cube will be to the ratio of the spheres of the tetrahedron as the consonances 1:2 and 1:3, which have been ascribed to the cube, are to 1:4 and 1:9, which are to be assigned to the tetrahedron, if this proportion is to be used. For these ratios, too, are as the squares of those consonances. And because 1:9 is not harmonic, 1:8 the proximate ratio takes its place in the tetrahedron. But by this proportion approximately 4:5 and 3:4 will go with the dodecahedral wedding. For as the ratio of the spheres of the cube is approximately the cube of the ratio of the dodecahedral, so too the cubic consonances 1:2 and 2:3 are approximately the cubes of the consonances 4:5 and 3:4. For 4:5 cubed is 64: 125, and 1:2 is 64:128. So 3:4 cubed is 27:64, and 1:3 is 27:81.

3. A SUMMARY OF ASTRONOMICAL DOCTRINE NECESSARY FOR SPECULATION INTO THE CELESTIAL HARMONIES

First of all, my readers should know that the ancient astronomical hypotheses of Ptolemy, in the fashion in which they have been unfolded in the *Theoricae* of Peurbach and by the other writers of epitomes, are to be completely removed from this discussion and cast out of the mind. For they do not convey the true layout of the bodies of the world and the polity of the movements.

Although I cannot do otherwise than to put solely Copernicus' opinion concerning the world in the place of those hypotheses and, if that were possible, to persuade everyone of it; but because the thing is still new among the mass of the intelligentsia (*apud vulgus studiosorum*), and the doctrine that the Earth is one of the planets and moves among the stars around a motionless sun sounds very absurd to the ears of most of them: therefore those who are shocked by the unfamiliarity of this opinion should know that these harmonical speculations are possible even with the hypotheses of Tycho Brahe—because that author holds, in common with Copernicus, everything else which pertains to the layout of the bodies and the tempering of the movements, and transfers solely the

Copernican annual movement of the Earth to the whole system of planetary spheres and to the Sun, which occupies the center of that system, in the opinion of both authors. For after this transference of movement it is nevertheless true that in Brahe the Earth occupies at any time the same place that Copernicus gives it, if not in the very vast and measureless region of the fixed stars, at least in the system of the planetary world. And accordingly, just as he who draws a circle on paper makes the writing-foot of the compass revolve, while he who fastens the paper or tablet to a turning lathe draws the same circle on the revolving tablet with the foot of the compass or stylus motionless; so too, in the case of Copernicus the Earth, by the real movement of its body, measures out a circle revolving midway between the circle of Mars on the outside and that of Venus on the inside; but in the case of Tycho Brahe the whole planetary system (wherein among the rest the circles of Mars and Venus are found) revolves like a tablet on a lathe and applies to the motionless Earth, or to the stylus on the lathe, the midspace between the circles of Mars and Venus; and it comes about from this movement of the system that the Earth within it, although remaining motionless, marks out the same circle around the sun and midway between Mars and Venus, which in Copernicus it marks out by the real movement of its body while the system is at rest. Therefore, since harmonic speculation considers the eccentric movements of the planets, as if seen from the Sun, you may easily understand that if any observer were stationed on a Sun as much in motion as you please, nevertheless for him the Earth, although at rest (as a

concession to Brahe), would seem to describe the annual circle midway between the planets and in an intermediate length of time. Wherefore, if there is any man of such feeble wit that he cannot grasp the movement of the Earth among the stars, nevertheless he can take pleasure in the most excellent spectacle of this most divine construction, if he applies to their image in the sun whatever he hears concerning the daily movements of the Earth in its eccentric—such an image as Tycho Brahe exhibits, with the Earth at rest.

And nevertheless the followers of the true Samian philosophy have no just cause to be jealous of sharing this delightful speculation with such persons, because their joy will be in many ways more perfect, as due to the consummate perfection of speculation, if they have accepted the immobility of the sun and the movement of the Earth.

Firstly [I], therefore, let my readers grasp that today it is absolutely certain among all astronomers that all the planets revolve around the Sun, with the exception of the Moon, which alone has the Earth as its center: the magnitude of the moon's sphere or orbit is not great enough for it to be delineated in this diagram in a just ratio to the rest. Therefore, to the other five planets, a sixth, the Earth, is added, which traces a sixth circle around the sun, whether by its own proper movement with the sun at rest, or motionless itself and with the whole planetary system revolving.

Secondly [II]: It is also certain that all the planets are eccentric, *i.e.*, they change their distances from the Sun, in such fashion that in one part of their circle they become farthest away from the Sun, and in the opposite part they come nearest to the Sun. In the accompanying diagram three circles apiece have been drawn for the single planets: none of them indicate the eccentric route of the planet itself; but the mean circle, such as *BE* in the case of Mars, is equal to the eccentric orbit, with respect to its longer diameter. But the orbit itself, such as *AD*, touches *AF*, the upper of the three, in one place *A*, and the lower circle *CD*, in the opposite place *D*. The circle *GH* made with dots and described through the center of the Sun indicates the route of the sun according to Tycho Brahe. And if the Sun moves on this route, then absolutely all the points in this whole planetary system here depicted advance upon an equal route, each upon his own. And with one point of it (namely, the center of the Sun) stationed at one point of its circle, as here at the lowest, absolutely each and every point of the system will be stationed at the lowest part of its circle. However, on account of the smallness of the space the three circles of Venus unite in one, contrary to my intention.

Kepler's calculation of the true orbit of Mars, from the relative different positions of the Earth.

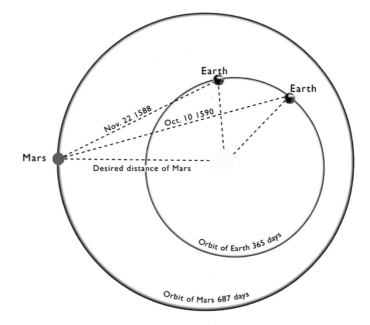

Thirdly [III]: Let the reader recall from my *Mysterium Cosmographicum*, which I published twenty-two years ago, that the number of the planets or circular routes around the sun was taken by the very wise Founder from the five regular solids, concerning which Euclid, so many ages ago, wrote his book which is called the *Elements* in that it is built up out of a series of propositions. But it has been made clear in the second book of this work that there cannot be more regular bodies, *i.e.*, that regular plane figures cannot fit together in a solid more than five times.

Fourthly [IV]: As regards the ratio of the planetary orbits, the ratio between two neighboring planetary orbits is always of such a magnitude that it is easily apparent that each and every one of them approaches the single ratio of the spheres of one of the five regular solids, namely, that of the sphere circumscribing to the sphere inscribed in the figure. Nevertheless it is not wholly equal, as I once dared to promise concerning the final perfection of astronomy. For, after completing the demonstration of the intervals from Brahe's observations, I discovered the following: if the angles of the cube are applied to the inmost circle of Saturn, the centers of the planes are approximately tangent to the middle circle of

Jupiter; and if the angles of the tetrahedron are placed against the inmost circle of Jupiter, the centers of the planes of the tetrahedron are approximately tangent to the outmost circle of Mars; thus if the angles of the octahedron are placed against any circle of Venus (for the total interval between the three has been very much reduced), the centers of the planes of the octahedron penetrate and descend deeply within the outmost circle of Mercury, but nonetheless do not reach as far as the middle circle of Mercury; and finally, closest of all to the ratios of the dodecahedral and icosahedral spheres—which ratios are equal to one another—are the ratios or intervals between the circles of Mars and the Earth, and the Earth and Venus; and those intervals are similarly equal, if we compute from the inmost circle of Mars to the middle circle of the Earth, but from the middle circle of the Earth to the middle circle of Venus. For the middle distance of the Earth is a mean proportional between the least distance of Mars and the middle distance of Venus. However, these two ratios between the planetary circles are still greater than the ratios of those two pairs of spheres in the figures, in such fashion that the centers of the dodecahedral planes are not tangent to the outmost circle of the Earth, and the centers of the icosahedral planes are not tangent to the outmost circle of Venus; nor, however, can this gap be filled by the semi-diameter of the lunar sphere, by adding it, on the upper side, to the greatest distance of the Earth and subtracting it, on the lower, from the least distance of the same. But I find a certain other ratio of figures—namely, if I take the augmented dodecahedron, to which I have given the name of echinus, (as being fashioned from twelve quinquangular stars and thereby very close to the five regular solids), if I take it, I say, and place its twelve points in the inmost circle of Mars, then the sides of the pentagons, which are the bases of the single rays or points, touch the middle circle of Venus. In short: the cube and the octahedron, which are consorts, do not penetrate their planetary spheres at all; the dodecahedron and the icosahedron, which are consorts, do not wholly reach to theirs, the tetrahedron exactly touches both: in the first case there is falling short; in the second, excess; and in the third, equality, with respect to the planetary intervals.

The world system determined from the geometry of the regular solids from Kepler's Harmonices Mundi Libri (Linz, 1619).

Wherefore it is clear that the very ratios of the planetary intervals from the Sun have not been taken from the regular solids alone. For the Creator, who is the very source of geometry and, as Plato wrote, "practices eternal geometry," does not stray from his own archetype. And indeed that very thing could be inferred from the fact that all the planets change their intervals throughout fixed periods of time, in such fashion that each has two marked intervals from the Sun, a greatest and a least; and a fourfold comparison of the intervals from the sun is possible between two planets: the comparison can be made between either the greatest, or the least, or the contrary intervals most remote from one another, or the contrary intervals nearest together. In this way the comparisons made two by two between neighboring planets are twenty in number, although on the contrary there are only five regular solids. But it is consonant that if the Creator had any concern for the ratio of the spheres in general, He would also have had concern for the ratio which exists between the varying intervals of the single planets specifically and the other. If we ponder that, we will comprehend that for setting up the that the concern is the same in both cases and the one is bound up with diameters and eccentricities conjointly, there is need of more principles, outside of the five regular solids.

Fifthly [V]: To arrive at the movements between which the consonances have been set up, once more I impress upon the reader that in the *Commentaries on Mars* I have demonstrated from the sure observations of Brahe that daily arcs, which are equal in one and the same eccentric circle, are not traversed with equal speed; but that these differing *delays in equal parts of the eccentric observe the ratio of their distances from the sun*, the source of movement; and conversely, that if equal times are assumed, namely, one natural day in both cases, the corresponding *true diurnal arcs of one eccentric orbit have to one another the ratio which is the inverse of the ratio of the two distances from the Sun*. Moreover, I demonstrated at the same time that *the planetary orbit is elliptical and the Sun, the source of movement, is at one of the foci of this ellipse; and so, when the planet has completed a quarter of its total circuit from its aphelion, then it is exactly at its mean distance from the sun, midway between its greatest distance at the aphelion and its least at the perihelion*. But from these two axioms it results *that the diurnal mean movement of the planet in its eccentric is the same as the true diurnal arc of its eccentric at those moments wherein the planet is at the end of the quadrant of the eccentric measured from the aphelion, although that true quadrant appears still smaller than the just quadrant.* Furthermore, it follows that *the sum of any two true diurnal eccentric arcs, one of which is at the same distance from the aphelion that the other is from the perihelion, is equal to the sum of the two mean diurnal arcs.* And as a consequence, *since the ratio of circles is the same as that of the diameters, the ratio of one mean diurnal arc to the sum of all the mean and equal arcs in the total circuit is the same as the ratio of the mean diurnal arc to the sum of all the true eccentric arcs, which are the same in number but unequal to one another.* And those things should first be known concerning the true diurnal arcs of the eccentric and the true movements, so that by means of them we may understand the movements which would be apparent if we were to suppose an eye at the sun.

Sixthly [VI]: But as regards the arcs which are apparent, as it were, from the Sun, it is known even from the ancient astronomy that, among true movements which are equal to one another, that movement which is farther distant from the center of the world (as being at the aphelion) will appear smaller to a beholder at that center, but the movement which is nearer (as being at the perihelion) will similarly appear greater. Therefore, since moreover the true diurnal arcs at the near distance are still greater, on account of the faster movement, and still smaller at the distant aphelion, on account of the slowness of the movement, I demonstrated in the *Commentaries on Mars that the ratio of the apparent diurnal arcs of one eccentric circle is fairly exactly the inverse ratio of the squares of their distances from the Sun.* For example, if the planet one day when it is at a distance from the sun of 10 parts, in any measure whatsoever, but on the opposite day, when it is at the perihelion, of 9 similar parts: it is certain that from the sun its apparent progress at the aphelion will be to its apparent progress at the perihelion, as 81:100.

But that is true with these provisos: First, that the eccentric arcs should not be great, lest they partake of distinct distances which are very different—*i.e.*, lest the distances of their termini from the apsides cause a perceptible variation; second, that the eccentricity should not be very great, for the greater its eccentricity (viz., the greater the arc becomes) the more the angle of its apparent movement increases beyond the measure of its approach to the Sun, by Theorem 8 of Euclid's *Optics*; none the less in small arcs even a great distance is of no moment, as I have remarked in my *Optics*, Chapter 11. But there is another reason why I make that admonition. For the eccentric arcs around the mean anomalies are viewed obliquely from the center of the Sun. This obliquity subtracts from the magnitude of the apparent movement, since conversely the arcs around the apsides are presented directly to an eye stationed as it were at the Sun. Therefore, when the eccentricity is very great, then the eccentricity takes away perceptibly from the ratio of the movements, if without any diminution we apply the mean diurnal movement to the mean distance, as if at the mean distance, it would appear to have the same magnitude which it does have—as will be apparent below in the case of Mercury. All these things are treated at greater length in Book V of the *Epitome of Copernican Astronomy*; but they have been mentioned here too because they have to do with the very terms of the celestial consonances, considered in themselves singly and separately.

Seventhly [VII]: If by chance anyone runs into those diurnal movements which are apparent to those gazing not as it were from the sun but from the Earth, with which movements Book VI of the *Epitome of Copernican Astronomy* deals, he should know that their rationale is plainly not considered in this business. Nor should it be, since the Earth is not the source of the planetary movements, nor can it be, since with respect to deception of sight they degenerate not only into mere quiet or apparent stations but even into retrogradation, in which way a whole infinity of ratios is assigned to all the planets, simultaneously and equally. Therefore, in order that we may hold for certain what sort of ratios of their own are constituted by the single real eccentric orbits (although these too are still apparent, as it were to one looking from the sun, the source of movement), first we must remove from those movements of their own this image of the adventitious annual movement common to all five, whether it arises from the movement of the Earth itself, according to Copernicus, or from the annual movement of the total system, according to Tycho Brahe, and the winnowed movements proper to each planet are to be presented to sight.

Eighthly [VIII]: So far we have dealt with the different delays or arcs of one and the same planet. Now we must also deal with the comparison of the movements of two planets. Here take note of the definitions of the terms which will be necessary for us. We give the name of *nearest apsides* of two planets to the perihelion of the upper and the aphelion of the lower, notwithstanding that they tend not towards the same region of the world but towards distinct and perhaps contrary regions. By *extreme movements* understand the slowest and the fastest of the whole planetary circuit; by *converging or converse extreme movements*, those which are at the nearest apsides of two planets—namely, at the perihelion of the upper planet and the aphelion of the lower; by *diverging or diverse*, those at the opposite apsides—namely, the aphelion of the upper and the perihelion of the lower. Therefore again, a certain part of my *Mysterium Cosmographicum*, which was suspended twenty-two years ago, because it was not yet clear, is to be completed

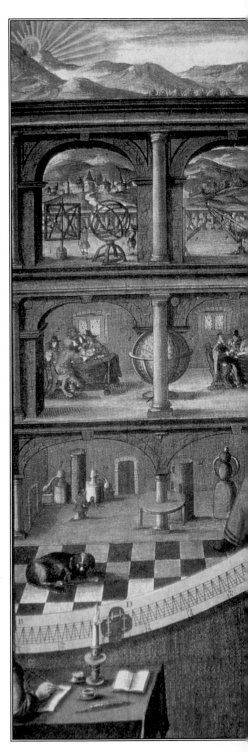

and herein inserted. For after finding the true intervals of the spheres by the observations of Tycho Brahe and continuous labour and much time, at last, at last the right ratio of the periodic times to the spheres though it was late, looked to the unskilled man, yet looked to him, and, after much time, came, and, if you want the exact time, was conceived mentally on the 8th of March in this year One Thousand Six Hundred and Eighteen but unfelicitously submitted to calculation and rejected as false, finally, summoned back on the 15th of May, with a fresh assault undertaken, outfought the darkness of my mind by the great proof afforded by my labor of seventeen years on Brahe's observations and meditation upon it uniting in one concord, in such fashion that I first believed I was dreaming and was presupposing the object of my search among the principles. But it is absolutely certain and exact that *the ratio which exists between the periodic times of any two planets is precisely the ratio of the 3/2th power of the mean distances*, i.e., *of the spheres themselves*; provided, however, that the arithmetic mean between both diameters of the elliptic orbit be slightly less than the longer diameter. And so if any one take the period, say, of the Earth, which is one year, and the period of Saturn, which is thirty years, and

CENTER

The mural in Tycho Brahe's Uranisborg observatory.

Tycho Braye's model.

extract the cube roots of this ratio and then square the ensuing ratio by squaring the cube roots, he will have as his numerical products the most just ratio of the distances of the Earth and Saturn from the Sun.[1] For the cube root of 1 is 1, and the square of it is 1; and the cube root of 30 is greater than 3, and therefore the square of it is greater than 9. And Saturn, at its mean distance from the Sun, is slightly higher than nine times the mean distance of the Earth from the Sun. Further on, in Chapter 9, the use of this theorem will be necessary for the demonstration of the eccentricities.

Ninthly [IX]: If now you wish to measure with the same yardstick, so to speak, the true daily journeys of each planet through the ether, two ratios are to be compounded—the ratio of the true (not the apparent) diurnal arcs of the eccentric, and the ratio of the mean intervals of each planet from the Sun (because that is the same as the ratio of the amplitude of the spheres), *i.e., the true diurnal arc of each planet is to be multiplied by the semidiameter of its sphere*: the products will be numbers fitted for investigating whether or not those journeys are in harmonic ratios.

Tenthly [X]: In order that you may truly know how great any one of these diurnal journeys appears to be to an eye stationed as it were at the Sun, although this same thing can be got immediately from the astronomy, nevertheless it will also be manifest if you multiply the ratio of the journeys by the inverse ratio not of the mean, but of the true intervals which exist at any position on the eccentrics: *Multiply the journey of the upper by the interval of the lower planet from the Sun, and conversely multiply the journey of the lower by the interval of the upper from the Sun.*

Eleventhly [XI]: And in the same way, if the apparent movements are given, at the aphelion of the one and at the perihelion of the other, or conversely or alternately, the ratios of the distances of the aphelion of the one to the perihelion of the other may be elicited. But where the mean movements must be known first, viz., the inverse ratio of the periodic times, wherefrom the ratio of the spheres is elicited by Article VIII above: then *if the mean proportional between the apparent movement of either one of its mean movement be taken, this mean proportional is to the semidiameter of its sphere* (which is already known) *as the mean movement is to the distance or interval sought.* Let the periodic times of two planets be 27 and 8. Therefore the ratio of the mean diurnal movement of the one to the other is 8 : 27. Therefore the semidiameters of their spheres will be as 9 to 4. For the cube root of 27 is 3, that of 8 is 2, and the squares of these roots, 3 and 2, are 9 and 4. Now let the apparent aphelial movement of the one be 2 and the perihelial movement of the other 331/3. The mean proportionals between the mean movements 8 and 27 and these apparent ones will be 4 and 30. Therefore if the mean proportional 4 gives the mean distance of 9 to the planet, then the mean movement of 8 gives an aphelial distance 18, which corresponds to the apparent movement 2; and if the other mean proportional 30 gives the other planet a mean distance of 4, then its mean movement of 27 will give it a perihelial interval of 33/5. I say, therefore, that the aphelial distance of the former is to the perihelial distance of the latter as 18 to 33/5. Hence it is clear that if the consonances between the extreme movements of two planets are found and the periodic times are established for both, the extreme and the mean distances are necessarily given, wherefore also the eccentricities.

Twelfthly [XII]: It is also possible, from the different extreme movements of one and the same planet, to find the *mean movement.* The mean movement is not exactly the arithmetic mean between the extreme movements, nor exactly the geometric mean, but it is as much less than the geometric mean as the geometric mean is less than the (arithmetic) mean between both means. Let the two extreme movements be 8 and 10: the mean movement will be less than 9, and also less than the square root of 80 by half the difference between 9 and the square root of 80. In this way, if the aphelial movement is 20 and the perihelial 24, the mean movement will be less than 22, even less than the square root of 480 by half the difference between that root and 22. There is use for this theorem in what follows.

Thirteenthly [XIII]: From the foregoing the following proposition is demonstrated, which is going to be very necessary for us: Just as the ratio of the mean movements of two planets is the inverse ratio of the 3/2th powers of the spheres, so the ratio of two apparent converging extreme movements always falls short of the ratio of the 3/2th powers of the intervals corresponding to those extreme movements; and in what ratio the product of the two ratios of the corresponding intervals to the two mean intervals or to the semidiameters of the two spheres falls short of the ratio of the square roots of the spheres, in that ratio does the ratio of the two extreme converging movements exceed the ratio of the corresponding intervals; but if that compound ratio were to exceed the ratio of the square roots of the spheres, then the ratio of the converging movements would be less than the ratio of their intervals.[2]

4. IN WHAT THINGS HAVING TO DO WITH THE PLANETARY MOVEMENTS MAVE THE HARMONIC CONSONANCES BEEN EXPRESSED BY THE CREATOR, AND IN WHAT WAY?

Accordingly, if the image of the retrogradation and stations is taken away and the proper movements of the planets in their real eccentric orbits are winnowed out, the following distinct things still remain in the planets: 1) The distances from the Sun. 2) The periodic times. 3) The diurnal eccentric arcs. 4) The diurnal delays in those arcs. 5) The angles at the Sun, and the diurnal area apparent to those as it were gazing from the Sun. And again, all of these things, with the exception of the periodic times, are variable in the total circuit, most variable at the mean longitudes, but least at the extremes, when, turning away from one extreme longitude, they begin to return to the opposite. Hence when the planet is lowest and nearest to the sun and thereby delays the least in one degree of its eccentric, and conversely in one day traverses the greatest diurnal arc of its eccentric and appears fastest from the Sun: then its movement remains for some time in this strength without perceptible variation, until, after passing the perihelion, the planet gradually begins to depart farther from the Sun in a straight line; at that same time it delays longer in the degrees of its eccentric circle; or, if you consider the movement of one day, on the following day it goes forward less and appears even more slow from the Sun until it has drawn close to the highest apsis and made its distance from the Sun very great: for then longest of all does it delay in one degree of its eccentric; or on the contrary in one day it traverses

*Mariner 10
passing by Mercury.*

its least arc and makes a much smaller apparent movement and the least of its total circuit.

Finally, all these things may be considered either as they exist in any one planet at different times or as they exist in different planets: Whence, by the assumption of an infinite amount of time, all the affects of the circuit of one planet can concur in the same moment of time with all the affects of the circuit of another planet and be compared, and then the total eccentrics, as compared with one another, have the same ratio as their semidiameters or mean intervals; but the arcs of two eccentrics, which are similar or designated by the same number (of degrees), nevertheless have their true lengths unequal in the ratio of their eccentrics. For example, one degree in the sphere of Saturn is approximately twice as long as one degree in the sphere of Jupiter. And conversely, the diurnal arcs of the eccentrics, as expressed in astronomical terms, do not exhibit the ratio of the true journeys which the globes complete in one day through the ether, because the single units in the wider circle of the upper planet denote a quarter part of the journey, but in the narrower circle of the lower planet a smaller part.

———————

7. THE UNIVERSAL CONSONANCES OF ALL SIX PLANETS, LIKE COMMON FOUR-PART COUNTERPOINT, CAN EXIST

But now, Urania, there is need for louder sound while I climb along the harmonic scale of the celestial movements to higher things where the true archetype of the fabric of the world is kept hidden. Follow after, ye modern musicians, and judge the thing according to your arts, which were unknown to antiquity. Nature, which is never not lavish of herself, after a lying-in of two thousand years, has finally brought you forth in these last generations, the first true images of the universe. By means of your concords of various voices, and through your ears, she has whispered to the human mind, the favorite daughter of God the Creator, how she exists in the innermost bosom.

(Shall I have committed a crime if I ask the single composers of this generation for some artistic motet instead of this epigraph? The Royal Psalter and the other Holy Books can supply a text suited for this. But alas for you! No more than six are in concord in the heavens. For the Moon sings here monody separately, like a dog sitting on the Earth. Compose the melody; I, in order that the book may progress, promise that I will watch carefully over the six parts. To him who more properly expresses the celestial music described in this work, Clio will give a garland, and Urania will betroth Venus his bride.)

It has been unfolded above what harmonic ratios two neighboring planets would embrace in their extreme movements. But it happens very rarely that two, especially the slowest, arrive at their extreme intervals at the same time; for example, the apsides of Saturn and Jupiter are about 81° apart. Accordingly, while this distance between them measures out the whole zodiac by definite twenty-year leaps,[3] eight hundred years pass

Robert Fludd's seventeenth-century drawing of the universe as a monochord. Many shared Kepler's view of a harmonic universe.

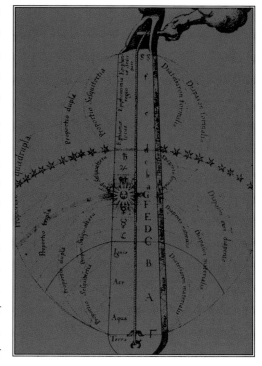

by, and nonetheless the leap which concludes the eighth century, does not carry precisely to the very apsides; and if it digresses much further, another eight hundred years must be awaited, that a more fortunate leap than that one may be sought; and the whole route must be repeated as many times as the measure of digression is contained in the length of one leap. Moreover, the other single pairs of planets have periods as that, although not so long. But meanwhile there occur also other consonances of two planets, between movements whereof not both are extremes but one or both are intermediate; and those consonances exist as it were in different tunings (*tensionibus*). For, because Saturn tends from G to *b*, and slightly further, and Jupiter from *b* to *d* and further; therefore between Jupiter and Saturn there can exist the following consonances, over and above the octave: the major and minor third and the perfect fourth, either one of the thirds through the tuning which maintains the amplitude of the remaining one, but the perfect fourth through the amplitude of a major whole tone. For there will be a perfect fourth not merely from G of Saturn to *cc* of Jupiter but also from A of Saturn to *dd* of Jupiter and through all the intermediates between the G and A of Saturn and the *cc* and *dd* of Jupiter. But the octave and the perfect fifth exist solely at the points of the apsides. But Mars, which got a greater interval as its own, received it in order that it should also make an octave with the upper planets through some amplitude of tuning. Mercury received an interval great enough for it to set up

almost all the consonances with all the planets within one of its periods, which is not longer than the space of three months. On the other hand, the Earth, and Venus much more so, on account of the smallness of their intervals, limit the consonances, which they form not merely with the others but with one another in especial, to visible fewness. But if three planets are to concord in one harmony, many periodic returns are to be awaited; nevertheless there are many consonances, so that they may so much the more easily take place, while each nearest consonance follows after its neighbor, and very often threefold consonances are seen to exist between Mars, the Earth, and Mercury. But the consonances of four planets now begin to be scattered throughout centuries, and those of five planets throughout thousands of years.

But that all six should be in concord has been fenced about by the longest intervals of time; and I do not know whether it is absolutely impossible for this to occur twice by precise evolving or whether that points to a certain beginning of time, from which every age of the world has flowed. But if only one sextuple can occur, or only one notable among many, indubitably that could be taken as a sign of the Creation.

But if only one sextuple harmony can occur, or only one notable one among many, indubitably that could be taken as a sign of the Creation. Therefore we must ask, in exactly how many forms are the movements of all six planets reduced to one common harmony? The method of inquiry is as follows: let us begin with the Earth and Venus, because these

Frontpiece of Franchino Gafari's Practica Musicae *(Milan, 1496).*

	Enneachor:on I	Enneach. II	Enneach. III	Enneach. IV
Diapafon Ditonus / Diapafon cum Touo / Diapafon / Heptachordon / Hexachordon / Diapente / Diateffaron / Ditonus / Tonus	Mundus Archetyp. DEVS	Mundus Sidereus Cœl.Emp.	Mundus Mineralis	Lapides
	Seraphim	Firmamentum	Salia, ftellę Minerales.	Aftrites
	Cherubim	♄ Nete	Plumbum	Topazius
	Troni	♃ Paranete	Æs	Amethiftus
	Dominationes	♂ Paramef.	Ferrum	Adamas
	Virtutes	☼ Mefe	Aurum	Pyropus
	Poteftates	♀ Lichanos	Stannum	Beryllus
	Principatus	☿ Parhypa.	Argentum Viuum	Achates Iafp is
	Archangeli	☽ Hypate	Argentum	Selenites Cryftallu
	Angeli	Ter.c ūEle. Proslamb.	Sulphur	Magnes

two planets do not make more than two consonances and (wherein the cause of this thing is comprehended) by means of very short intensifications of the movements. Therefore let us set up two, as it were, skeletal outlines of harmonies, each skeletal outline determined by the two extreme numbers wherewith the limits of the tunings are designated, and let us search out what fits in with them from the variety of movements granted to each planet.

———

Enneach: V	Enneach. VI	Enneach VII	Enneach: VIII	Enneach. IX	Enneach. X
...anta	Arbores	Aquatilia	Volucria	Quadrupepedia	Colores varij
...rbæ & ...r.ſtell.	Frutices Bacciferę	Piſces ſtellares	Gallina Pharaonis	Pardus	Diuerſi Colores
...leborus	Cypreſſus	Tynnus	Bubo	Aſinus, Vrſus	Fuſcus
...tonica	Citrus	Acipenſer	Aquila	Elephas	Roſcus
...ſynthiũ	Quercus	Pſyphias	Falco Accipiter	Lupus	Flammeus
...liotrop ...ium	Łotus, Laurus	Delphinus	Gallus	Leo	Aureus
...yrium	Myrtus	Truta	Cygnus Columba	Ceruus	Viridis
...æonia	Maluſpunica	Caſtor	Pſittacus	Canis	Cæruleus
...unaria	Colutea	Oſtrea	Anates Anſeres	Ælurus	Candidus
...amina	Frutices	Anguilla	Struthio camelus	Inſecta	Niger

The universe as a harmonious arrangement based on the number 9. Athanasius Kircher's Musurgia Universalis (Rome, 1650).

THE END

Isaac Newton (1642-1727)

HIS LIFE AND WORK

On February 5, 1676, Isaac Newton penned a letter to his bitter enemy, Robert Hooke, which contained the sentence, "If I have seen farther, it is by standing on the shoulders of giants." Often described as Newton's nod to the scientific discoveries of Copernicus, Galileo, and Kepler before him, it has become one of the most famous quotes in the history of science. Indeed, Newton did recognize the contributions of those men, some publicly and others in private writings. But in his letter to Hooke, Newton was referring to optical theories, specifically the study of the phenomena of thin plates, to which Hooke and Renè Descartes had made significant contributions.

Some scholars have interpreted the sentence as a thinly veiled insult to Hooke, whose crooked posture and short stature made him anything but a giant, especially in the eyes of the extremely vindictive Newton. Yet despite their feuds, Newton did appear to humbly acknowledge the noteworthy research in optics of both Hooke and Descartes, adopting a more conciliatory tone at the end of the letter.

Isaac Newton is considered the father of the study of infinitesimal calculus, mechanics, and planetary motion, and the theory of light and color. But he secured his place in history by formulating gravitational force and defining the laws of motion and attraction in his landmark work, *Mathematical Principles of Natural Philosophy* (*Philosophiae Naturalis Principia Mathematica*) generally known as Principia. There Newton fused the scientific contributions of Copernicus, Galileo, Kepler, and others into a dynamic new symphony. *Principia*, the first book on theoretical physics, is roundly regarded as the most important work in the history of science and the scientific foundation of the modern worldview.

Newton wrote the three books that form *Principia* in just eighteen months and, astonishingly, between severe emotional breakdowns—likely compounded by his competition with Hooke. He even went to such vindictive lengths as to remove from the book all references to Hooke's work, yet his hatred for his fellow scientist may have been the very inspiration for *Principia*.

The slightest criticism of his work, even if cloaked in lavish praise, often sent Newton into dark withdrawal for months or years. This trait revealed itself early in Newton's life and has led some to wonder what other questions Newton might have answered had he not been obsessed with settling personal feuds. Others have speculated that Newton's scientific discoveries and achievements were the result of his vindictive obsessions and might not have been possible had he been less arrogant.

As a young boy, Isaac Newton asked himself the questions that had long mystified humanity, and then went on to answer many of them. It was the beginning of a life full of discovery, despite some anguishing first steps. Isaac Newton was born in the English industrial town of Woolsthorpe, Lincolnshire, on Christmas Day of 1642, the same year in which Galileo died. His mother did not expect him to live long, as he was born very prematurely; he would later describe himself as having been so small at birth he could fit into a quart pot. Newton's yeoman father, also named Isaac, had died three month's earlier, and when Newton reached two years of age, his mother, Hannah Ayscough, remarried, wedding Barnabas Smith, a rich clergyman from North Witham.

Apparently there was no place in the new Smith family for the young Newton, and he was placed in the care of his grandmother, Margery Ayscough. The specter of this abandonment, coupled with the tragedy of never having known his father, haunted Newton for the rest of his life. He despised his stepfather; in journal entries for 1662 Newton, examining his sins, recalled "threatening my father and mother Smith to burne them and the house over them."

Newton at 12.

Much like his adulthood, Newton's childhood was filled with episodes of harsh, vindictive attacks, not only against perceived enemies but against friends and family as well. He also displayed the kind of curiosity early on that would define his life's achievements, taking an interest in mechanical models and architectural drawing. Newton spent countless hours building clocks, flaming kites, sundials, and miniature mills (powered by mice) as well as drawing elaborate sketches of animals and ships. At the age of five he attended schools at Skillington and Stoke but was considered one of the poorest students, receiving comments in teachers' reports such as "inattentive" and "idle." Despite his curiosity and demonstrable passion for learning, he was unable to apply himself to schoolwork.

By the time Newton reached the age of ten, Barnabas Smith had passed away and Hannah had come into a considerable sum from Smith's estate. Isaac and his grandmother began living with Hannah, a half-brother, and two half-sisters. Because his work at school was uninspiring, Hannah decided that Isaac would be better off managing the farm and estate, and she pulled him out of the Free Grammar School in Grantham. Unfortunately for her, Newton had even less skill or interest in managing the family estate than he had in schoolwork. Hannah's brother, William, a clergyman, decided that it would be best for the family if the absent-minded Isaac returned to school to finish his education.

This time, Newton lived with the headmaster of the Free Grammar School, John Stokes, and he seemed to turn a corner in his education. One story has it that a blow to the head, administered by a schoolyard bully, somehow enlightened him, enabling the young Newton to reverse the negative course of his educational promise. Now demonstrating intellectual aptitude and curiosity, Newton began preparing for further study at a university. He decided to attend Trinity College, his uncle William's alma mater, at Cambridge University.

At Trinity, Newton became a subsizar, receiving an allowance toward the cost of his education in exchange for performing various chores such as waiting tables and cleaning rooms for the faculty. But by 1664 he was elected scholar, which guaranteed him financial support and freed him

Cartoon of the story that Newton discovered gravity when he was struck on the head by a falling apple.

from menial duties. When the university closed because of the bubonic plague in 1665, Newton retreated to Lincolnshire. In the eighteen months he spent at home during the plague he devoted himself to mechanics and mathematics, and began to concentrate on optics and gravitation. This "*annus mirabilis*" (miraculous year), as Newton called it, was one of the most productive and fruitful periods of his life. It is also around this time that an apple, according to legend, fell onto Newton's head, awakening him from a nap under a tree and spurring him on to

Newton conducting experiments
with a prism in his room at
Trinity College.

define the laws of gravity. However far-fetched the tale, Newton himself wrote that a falling apple had "occasioned" his foray into gravitational contemplation, and he is believed to have performed his pendulum experiments then. "I was in the prime of my age for invention," Newton later recalled, "and minded Mathematicks and Philosophy more than at any time since."

When he returned to Cambridge, Newton studied the philosophy of Aristotle and Descartes, as well as the science of Thomas Hobbes and

Robert Boyle. He was taken by the mechanics of Copernicus and Galileo's astronomy, in addition to Kepler's optics. Around this time, Newton began his prism experiments in light refraction and dispersion, possibly in his room at Trinity or at home in Woolsthorpe. A development at the university that clearly had a profound influence on Newton's future—was the arrival of Isaac Barrow, who had been named the Lucasian Professor of Mathematics. Barrow recognized Newton's extraordinary mathematical talents, and when he resigned his professorship in 1669 to pursue theology he recommended the twenty-seven-year old Newton as his replacement.

Newton's first studies as Lucasian Professor centered in the field of optics. He set out to prove that white light was composed of a mixture of various types of light, each producing a different color of the spectrum when refracted by a prism. His series of elaborate and precise experiments to prove that light was composed of minute particles drew the ire of scientists such as Hooke, who contended that light traveled in waves. Hooke challenged Newton to offer further proof of his eccentric optical theories. Newton's way of responding was one he did not outgrow as he matured. He withdrew, set out to humiliate Hooke at every opportunity, and refused to publish his book, *Opticks*, until after Hooke's death in 1703.

Early in his tenure as Lucasian Professor, Newton was well along in his study of pure mathematics, but he shared his work with very few of his colleagues. Already by 1666, he had discovered general methods of solving problems of curvature—what he termed "theories of fluxions and inverse fluxions." The discovery set off a dramatic feud with supporters of the German mathematician and philosopher Gottfried Wilhelm Leibniz, who more than a decade later published his findings on differential and integral calculus. Both men arrived at roughly the same mathematical principles, but Leibniz published his work before Newton. Newton's supporters claimed that Leibniz had seen the Lucasian Professor's papers years before, and a heated argument between the two camps, known as the Calculus Priority Dispute, did not end until Leibniz died in 1716. Newton's vicious attacks which often spilled over to touch

The Goddess Artemis holding an image of Newton.

on views about God and the universe, as well as his accusations of pla-giarism, left Leibniz impoverished and disgraced.

Most historians of science believe that the two men in fact arrived at their ideas independently and that the dispute was pointless. Newton's vitriolic aggression toward Leibniz took a physical and emotional toll on Newton as well. He soon found himself involved in another battle, this time over his theory of color, and in 1678 he suffered a severe mental breakdown. The next year, his mother, Hannah passed away, and Newton began to distance himself from others. In secret, he delved into alchemy, a field widely regarded already in Newton's time as fruitless. This episode in the scientist's life has been a source of embarrassment to many Newton scholars. Only long after Newton died did it become apparent that his interest in chemical experiments was related to his later research in celestial mechanics and gravitation.

Newton had already begun forming theories about motion by 1666, but he was as yet unable to adequately explain the mechanics of circular motion. Some fifty years earlier, the German mathematician and astronomer Johannes Kepler had proposed three laws of planetary motion, which accurately described how the planets moved in relation to the sun, but he could not explain why the planets moved as they did. The closest Kepler came to understanding the forces involved was to say that the sun and the planets were "magnetically" related.

Newton set out to discover the cause of the planets' elliptical orbits. By applying his own law of centrifugal force to Kepler's third law of planetary motion (the law of harmonies) he deduced the inverse-square law, which states that the force of gravity between any two objects is inversely proportional to the square of the distance between the object's centers. Newton was thereby coming to recognize that gravitation is universal—that one and the same force causes an apple to fall to the ground and the Moon to race around the Earth. He then set out to test the inverse-square relation against known data. He accepted Galileo's estimate that the Moon is sixty earth radii from the Earth, but the inac-curacy of his own estimate of the Earth's diameter made it impossible to complete the test to his satisfaction. Ironically, it was an exchange of

letters in 1679 with his old adversary Hooke that renewed his interest in the problem. This time, he turned his attention to Kepler's second law, the law of equal areas, which Newton was able to prove held true because of centripetal force. Hooke, too, was attempting to explain the planetary orbits, and some of his letters on that account were of particular interest to Newton.

At an infamous gathering in 1684, three members of the Royal Society—Robert Hooke, Edmond Halley, and Christopher Wren, the noted architect of St. Paul's Cathedral—engaged in a heated discussion about the inverse-square relation governing the motions of the planets. In the early 1670s, the talk in the coffeehouses of London and other intellectual centers was that gravity emanated from the Sun in all directions and fell off at a rate inverse to the square of the distance, thus

William Blake's 1795 color print of Newton.

Newton's Principia.

becoming more and more diluted over the surface of the sphere as that surface expands. The 1684 meeting was, in effect, the birth of *Principia*. Hooke declared that he had derived from Kepler's law of ellipses the proof that gravity was an emanating force, but would withhold it from Halley and Wren until he was ready to make it public. Furious, Halley went to Cambridge, told Newton Hooke's claim, and proposed the following problem. "What would be the form of a planet's orbit about the Sun if it were drawn towards the Sun by a force that varied inversely as the square of the distance?" Newton's response was staggering. "It would be an ellipse," he answered immediately, and then told Halley that he had solved the problem four years earlier but had misplaced the proof in his office.

At Halley's request, Newton spent three months reconstituting and improving the proof. Then, in a burst of energy sustained for eighteen months, during which he was so caught up in his work that he often forgot to eat, he further developed these ideas until their presentation filled three volumes. Newton chose to title the work *Philosophiae Naturalis Principia Mathematica*, in deliberate contrast with Descartes' *Principia Philosophiae*. The three books of Newton's *Principia* provided the link between Kepler's laws and the physical world. Halley reacted with "joy and amazement" to Newton's discoveries. To Halley, it seemed the Lucasian Professor had succeeded where all others had failed, and he personally financed publication of the massive work as a masterpiece and a gift to humanity.

Where Galileo had shown that objects were "pulled" toward the center of the Earth, Newton was able to prove that this same force, gravity, affected the orbits of the planets. He was also familiar with Galileo's work on the motion of projectiles, and he asserted that the Moon's orbit around the Earth adhered to the same principles. Newton demonstrated that gravity explained and predict the Moon's motions as well as the rising and falling of the tides on Earth. Book 1 of *Principia* encompasses Newton's three laws of motion:

1. Every body perseveres in its state of resting, or uniformly moving in a right line, unless it is compelled to change that state by forces impressed upon it.

2. The change of motion is proportional to the motive force impressed; and is made in the direction of the right line in which that force is impressed.

3. To every action there is always opposed an equal reaction; or, the mutual actions of two bodies upon each other are always equal, and directed to contrary directions.

Book 2 began for Newton as something of an afterthought to Book 1; it was not included in the original outline of the work. It is essentially a treatise on fluid mechanics, and it allowed Newton room to display his mathematical ingenuity. Toward the end of the book, Newton concludes that the vortices invoked by Descartes to explain the motions of planets do not hold up to scrutiny, for the motions could be performed in free space without vortices. How that is so, Newton wrote, "may be understood by the first Book; and I shall now more fully treat of it in the following Book."

In Book 3, subtitled *System of the World,* by applying the laws of motion from book 1 to the physical world Newton concluded that

"there is a power of gravity tending to all bodies, proportional to the several quantities of matter which they contain." He thus demonstrated that his law of universal gravitation could explain the motions of the six known planets, as well as moons, comets, equinoxes, and tides. The law states that all matter is mutually attracted with a force directly proportional to the product of their masses and inversely proportional to the square of the distance between them. Newton, by a single set of laws, had united the Earth with all that could be seen in the skies. In the first two "Rules of Reasoning" from Book 3, Newton wrote:

We are to admit no more causes of natural things than such as are both true and sufficient to explain their appearances. Therefore, to the same natural effects we must, as far as possible, assign the same causes.

It is the second rule that actually unifies heaven and earth. An Aristotelian would have asserted that heavenly motions and terrestrial motions are manifestly not the same natural effects and that Newton's second rule could not, therefore, be applied. Newton saw things otherwise.

Principia was moderately praised upon its publication in 1687, but only about five hundred copies of the first edition were printed. However, Newton's nemesis, Robert Hooke, had threatened to spoil any coronation Newton might have enjoyed. After Book 2 appeared, Hooke publicly claimed that the letters he had written in 1679 had provided scientific ideas that were vital to Newton's discoveries. His claims, though not without merit, were abhorrent to Newton, who vowed to delay or even abandon publication of Book 3. Newton ultimately relented and published the final book of *Principia*, but not before painstakingly removing from it every mention of Hooke's name.

Newton's hatred for Hooke consumed him for years afterward. In 1693, he suffered yet another nervous breakdown and retired from research. He withdrew from the Royal Society until Hooke's death in 1703, then was elected its president and reelected each year until his own death in 1727. He also withheld publication of *Opticks*, his important

OPPOSITE PAGE

Eighteenth-century cartoon mocking Newton's theories on gravity.

study of light and color that would become his most widely read work, until after Hooke was dead.

Newton began the eighteenth century in a government post as warden of the Royal Mint, where he utilized his work in alchemy to determine methods for reestablishing the integrity of the English currency. As president of the Royal Society, he continued to battle perceived enemies with inexorable determination, in particular carrying on his longstanding feud with Leibniz over their competing claims to have invented calculus. He was knighted by Queen Anne in 1705, and lived to see publication of the second and third editions of *Principia*.

Isaac Newton died in March 1727, after bouts of pulmonary inflammation and gout. As was his wish, Newton had no rival in the field of science. The man who apparently formed no romantic attachments with women (some historians have speculated on possible relationships with men, such as the Swiss natural philosopher Nicolas Fatio de Duillier) cannot, however, be accused of a lack of passion for his work. The poet Alexander Pope, a contemporary of Newton's, most elegantly described the great thinker's gift to humanity:

> *Nature and Nature's laws lay hid in night:*
> *God said, "Let Newton be! and all was light."*

For all the petty arguments and undeniable arrogance that marked his life, toward its end Isaac Newton was remarkably poignant in assessing his accomplishments: "I do not know how I may appear to the world, but to myself I seem to have been only like a boy, playing on the seashore, and diverting myself, in now and then finding a smoother pebble or prettier shell than ordinary, whilst the great ocean of truth lay all undiscovered before me."

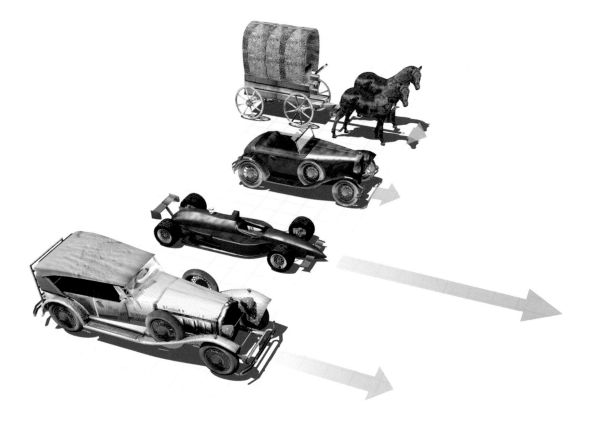

Newton's second law states that a body will accelerate or change speed at a rate that is proportional to its force. The acceleration is smaller the greater the mass of the body. A car with a 250-brake-horsepower engine has a greater acceleration than one with only twenty-five bhp. However, a car weighing twice as much will accelerate at half the rate of the smaller and lighter car.

PRINCIPIA

THE MATHEMATICAL PRINCIPLES OF NATURAL PHILOSOPHY

AXIOMS, OR LAWS OF MOTION

LAW I. EVERY BODY PERSERVERES IN ITS STATE OF REST, OR OF UNIFORM MOTION IN A RIGHT LINE, UNLESS IT IS COMPELLED TO CHANGE THAT STATE BY FORCES IMPRESSED THERON.

Projectiles persevere in their motions, so far as they are not retarded by the resistance of the air, or impelled downwards by the force of gravity A top, whose parts by their cohesion are perpetually drawn aside from rectilinear motions, does not cease its rotation, otherwise than as it is retarded by the air. The greater bodies of the planets and comets, meeting with less resistance in more free spaces, preserve their motions both progressive and circular for a much longer time.

LAW II. THE ALTERATION OF MOTION IS EVER PROPORTIONAL TO THE MOTIVE FORCE IMPRESSED; AND IS MADE IN THE DIRECTION OF THE RIGHT LINE IN WHICH THAT FORCE IS IMPRESSED.

If any force generates a motion, a double force will generate double the motion, a triple force triple the motion, whether that force be impressed altogether and at once, or gradually and successively. And this motion (being always directed the same way with the generating force), if the body moved before, is added to or subducted from the former motion, according as they directly conspire with or are directly contrary to each other; or obliquely joined, when they are oblique, so as to produce a new motion compounded from the determination of both.

LAW III. TO EVERY ACTION THERE IS ALWAYS OPPOSED AN EQUAL REACTION: OR THE MUTUAL ACTIONS OF TWO BODIES UPON EACH OTHER ARE ALWAYS EQUAL, AND DIRECTED TO CONTRARY PARTS.

Whatever draws or presses another is as much drawn or pressed by that other. If you press a stone with your finger, the finger is also pressed by the stone. If a horse draws a stone tied to a rope, the horse (if I may so say) will be equally drawn back towards the stone: for the distended rope, by the same endeavor to relax or unbend itself, will draw the horse as much towards the stone, as it does the stone towards the horse, and will obstruct the progress of the one as much as it advances that of the other.

If a body impinge upon another, and by its force change the motion of the other, that body also (because of the equality of the mutual pressure) will undergo an equal change, in its own motion, towards the contrary part. The changes made by these actions are equal, not in the velocities but in the motions of bodies; that is to say, if the bodies are not hindered by any other impediments. For, because the motions are equally changed, the changes of the velocities made towards contrary parts are reciprocally proportional to the bodies. This law takes place also in attractions, as will be proved in the next scholium.

COROLLARY I. A BODY BY TWO FORCES CONJOINED WILL DESCRIBE THE DIAGONAL OF A PARALLELOGRAM, IN THE SAME TIME THAT IT WOULD DESCRIBE THE SIDES, BYT THOSE FORCES APART.

If a body in a given time, by the force M impressed apart in the place A, should with a uniform motion be carried from A to B; and by the force N impressed apart in the same place, should be carried from A to C; complete the parallelogram ABCD, and, by both forces acting together, it will in the same time be carried in the diagonal from A to D. For since the force N acts in the direction of the line AC, parallel to BD, this force (by the second law) will not at all alter the velocity generated by the other force M, by which the body is carried towards the line BD. The body therefore will arrive at the line BD in the same time, whether the force N be impressed or not; and therefore at the end of that time it will be found somewhere in the line BD. By the same argument, at the end of the same time it will be found somewhere in the line CD. Therefore it will be found in the point D, where both lines meet. But it will move in a right line from A to D, by Law I.

COROLLARY II. AND HENCE IS EXPLAINED THE COMPOSTION OF ANY ONE DIRECT FORCE AD, OUT OF ANY TWO OBLIQUE FORCES AC AND CD; AND, ON THE CONTRARY, THE RESOLUTION OF ANY ONE DIRECT FORCE AND INTO TWO OBLIQUE FORCES AC AND CD: WHICH COMPOSITION AND RESOLUTION ARE ABUNDANTLY CONFIRMED FROM MECHANICS.

As if the unequal radii OM and ON drawn from the center O of any wheel, should sustain the weights A and P by the cords MA and NP; and the forces of those weights to move the wheel were required. Through the center O draw the right line KOL, meeting the cords perpendicularly in K and L; and from the center O, with OL the greater of the distances OK and OL, describe a circle, meeting the cord MA in D: and drawing OD, make AC parallel and DC perpendicular thereto. Now, it being indifferent whether the points K, L, D, of the cords be fixed to the plane of the wheel or not, the weights will have the same effect whether they are suspended from the points K and L, or from D and L. Let the whole force of the weight A be represented by the line AD, and let it be resolved into the forces AC and CD; of which the force AC, drawing the radius OD directly from the center, will have no effect to move the wheel: but the other force DC, drawing the radius DO perpendicularly, will have the same effect as if it drew perpendicularly the radius OL equal to OD; that is, it will have the same effect as the weight P, if that weight is to the weight A as the force DC is to the force DA; that is (because of the similar triangles ADC, DOK), as OK to OD or OL. Therefore the weights A and P, which are reciprocally as the radii OK and OL that lie in the same right line, will be equipollent, and so remain in equilibrio; which is the well known property of the balance, the lever, and the wheel. If either weight is greater than in this ratio, its force to move the wheel will be so much greater.

If the weight p, equal to the weight P, is partly suspended by the cord Np, partly sustained by the oblique plane pG; draw pH, NH, the former

perpendicular to the horizon, the latter to the plane pG; and if the force of the weight p tending downwards is represented by the line pH, it may be resolved into the forces pN, HN. If there was any plane pQ, perpendicular to the cord pN, cutting the other plane pG in a line parallel to the horizon, and the weight p was supported only by those planes pQ, pG, it would press those planes perpendicularly with the forces pN, HN; to wit, the plane pQ with the force pN, and the plane pG with the force

HN. And therefore if the plane pQ was taken away, so that the weight might stretch the cord, because the cord, now sustaining the weight, supplies the place of the plane that was removed, it will be strained by the same force N which pressed upon the plane before. Therefore, the tension of this oblique cord pN will be to that of the other perpendicular cord PN as pN to pH. And therefore if the weight p is to the weight A in a ratio compounded of the reciprocal ratio of the least distances of the cords PN, AM, from the center of the wheel, and of the direct ratio of pH to pN, the weights will have the same effect towards moving the wheel and will therefore sustain each other; as any one may find by experiment.

But the weight p pressing upon those two oblique planes may be considered as a wedge between the two internal surfaces of a body split by it; and hence the forces of the wedge and the mallet may be determined; for because the force with which the weight p presses the plane pQ is to the force with which the same, whether by its own gravity, or by the blow of a mallet, is impelled in the direction of the line pH towards both the planes, as pH to pH; and to the force with which it presses the other plane pG, as pN to NH. And thus the force of the screw may be deduced from a like resolution of forces; it being no other than a wedge impelled with the force of a lever. Therefore the use of this Corollary spreads far and wide, and by that diffusive extent the truth thereof is farther confirmed. For on what has been said depends the whole doctrine of mechanics variously demonstrated by different authors. For from hence are easily deduced the forces of machines, which are compounded of wheels, pullies, levers, cords, and weights, ascending directly or obliquely, and other mechanical powers; as also the force of the tendons to move the bones of animals.

COROLLARY III. THE QUANTITY OF MOTION, WHICH IS COLLECTED BY TAKING THE SUM OF THE MOTIONS DIRECTED TOWARDS THE SAME PARTS, AND THE DIFFERENCE OF THOSE THAT ARE DIRECTED TO CONTRARY PARTS, SUFFERS NO CHANGE FROM THE ACTION OF BODIES AMONG THEMSELVES.

OPPOSITE PAGE

Newton's diagram of the fist reflecting telescope he built in 1668.

For action and its opposite reaction are equal, by Law III, and therefore, by Law II, they produce in the motions equal changes towards opposite parts. Therefore if the motions are directed towards the same parts, whatever is added to the motion of the preceding body will be subducted from the motion of that which follows; so that the sum win be the same as before. If the bodies meet, with contrary motions, there will be an equal deduction from the motions of both; and therefore the difference of the motions directed towards opposite parts will remain the same.

Thus if a spherical body A with two parts of velocity is triple of a spherical body B which follows in the same right line with ten parts of velocity, the motion of A will be to that of B as 6 to 10. Suppose, then, their motions to be of 6 parts and of 10 parts, and the sum will be 16 parts. Therefore, upon the meeting of the bodies, if A acquire 3, 4, or 5 parts of motion, B will lose as many; and therefore after reflection A will proceed with 9, 10, or 11 parts, and B with 7, 6, or 5 parts; the sum remaining always of 16 parts as before. If the body A acquire 9, 10, 11, or 12 parts of motion, and therefore after meeting proceed with 15, 16, 17, or 18 parts, the body B, losing so many parts as A has got, will either proceed with 1 part, having lost 9, or stop and remain at rest, as having lost its whole progressive motion of 10 parts; or it will go back with 1 part, having not only lost its whole motion, but (if I may so say) one part more; or it will go back with 2 parts, because a progressive motion of 12 parts is taken off. And so the sums of the conspiring motions 15+1 or 16+0, and the differences of the contrary motions 17-1 and 18-2, will always be equal to 16 parts, as they were before the meeting and reflection of the bodies. But, the motions being known with which the bodies proceed after reflection, the velocity of either will be also known, by taking the velocity after to the velocity before reflection, as the motion after is to the motion before. As in the last case, where the motion of the body A was of 6 parts before reflection and of 18 parts after, and the velocity was of 2 parts before reflection, the velocity thereof after reflection will be found to be of 6 parts; by saying, as the 6 parts of motion, before to 18 parts after, so are 2 parts of velocity before reflection to 6 parts after.

But if the bodies are either not spherical, or, moving in different right lines, impinge obliquely one upon the other, and their motions after reflection are required, in those cases we are first to determine the position of the plane that touches the concurring bodies in the point of concourse; then the motion of each body (by Corol. II) is to be resolved into two, one perpendicular to that plane, and the other parallel to it. This done, because the bodies act upon each other in the direction of a line perpendicular to this plane, the parallel motions are to be retained the same after reflection as before; and to the perpendicular motions we are to assign equal changes towards the contrary parts; in such manner that the sum of the conspiring and the difference of the contrary motions may remain the same as before. From such kind of reflections also sometimes arise the circular motions of bodies about their own centers. But these are cases which I do not consider in what follows; and it would be too tedious to demonstrate every particular that relates to this subject.

COROLLARY IV. THE COMMON CENTER OF GRAVITY OF TWO OR MORE BODIES DOES NOT ALTER ITS STATE OF MOTION OR REST BY THE ACTIONS OF THE BODIES AMONG THEMSELVES: AND THEREFORE THE COMMON CENTER OF GRAVITY OF ALL BODIES ACTING UPON EACH OTHER (EXCLUDING OUTWARD ACTIONS AND IMPEDIMENTS) IS EITHER AT REST, OR MOVES UNIFORMLY IN A RIGHT LINE.

For if two points proceed with an uniform motion in right lines, and their distance be divided in a given ratio, the dividing point will be either at rest, or proceed uniformly in a right line. This is demonstrated hereafter in Lem. XXIII and its Corol., when the points are moved in the same plane; and by a like way of arguing, it may be demonstrated when the points are not moved in the same plane. Therefore if any number of bodies move uniformly in right lines, the common center of gravity of any two of them is either at rest, or proceeds uniformly in a right line; because the line which connects the centers of those two bodies so moving is divided at that common center in a given ratio. In like manner the common center of those two and that of a third body will be either at rest or moving uniformly in a right line because at that center the

169

distance between the common center of the two bodies, and the center of this last, is divided in a given ratio. In like manner the common center of these three, and of a fourth body, is either at rest, or moves uniformly in a right line; because the distance between the common center of the three bodies, and the center of the fourth is there also divided in a given ratio, and so on *ad infinitum*. Therefore, in a system of bodies where there is neither any mutual action among themselves, nor any foreign force impressed upon them from without, and which consequently move uniformly in right lines, the common center of gravity of them all is either at rest or moves uniformly forward in a right line.

Moreover, in a system of two bodies mutually acting upon each other, since the distances between their centers and the common center of gravity of both are reciprocally as the bodies, the relative motions of those bodies, whether of approaching to or of receding from that center, will be equal among themselves. Therefore since the changes which happen to motions are equal and directed to contrary parts, the common center of those bodies, by their mutual action between themselves, is neither promoted nor retarded, nor suffers any change as to its state of motion or rest. But in a system of several bodies, because the common center of gravity of any two acting mutually upon each other suffers no change in its state by that action; and much less the common center of gravity of the others with which that action does not intervene: but the distance between those two centers is divided by the common center of gravity of all the bodies into parts reciprocally proportional to the total sums of those bodies whose centers they are: and therefore while those two centers retain their state of motion or rest, the common center of all does also retain its state: it is manifest that the common center of all never suffers any change in the state of its motion or rest from the actions of any two bodies between themselves. But in such a system all the actions of the bodies among themselves either happen between two bodies, or are composed of actions interchanged between some two bodies; and therefore they do never produce any alteration in the common center of all as to its state of motion or rest. Wherefore since that center, when the bodies do not act mutually one upon another, either is at rest or moves

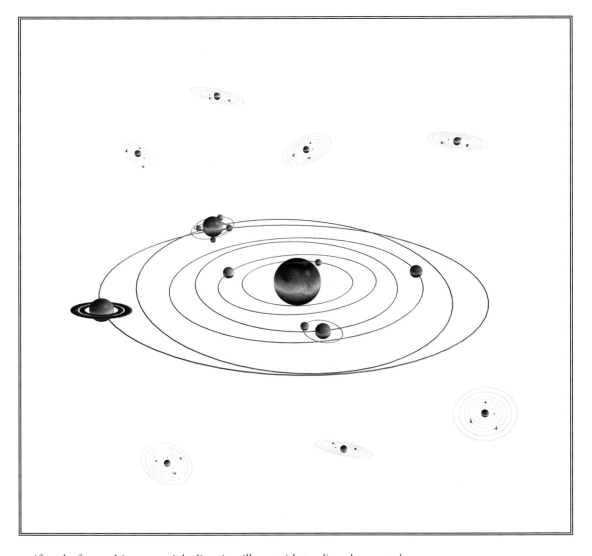

uniformly forward in some right line, it will, notwithstanding the mutual actions of the bodies among themselves, always persevere in its state, either of rest, or of proceeding uniformly in a right line, unless it is forced out of this state by the action of some power impressed from without upon the whole system. And therefore the same law takes place in a system consisting of many bodies as in one single body, with regard to their persevering in their state of motion or of rest. For the progressive motion, whether of one single body, or of a whole system of bodies, is always to be estimated from the motion of the center of gravity.

COROLLARY V. THE MOTIONS OF BODIES INCLUDED IN A GIVEN SPACE ARE THE SAME AMONG THEMSELVES, WHETHER THAT SPACE IS AT REST, OR MOVE UNIFORMLY FORWARDS IN A RIGHT LINE WITHOUT ANY CIRCULAR MOTION.

For the differences of the motions tending towards the same parts, and the sums of those that tend towards contrary parts, are, at first (by supposition), in both cases the same; and it is from those sums and differences that the collisions and impulses do arise with which the bodies mutually impinge one upon another. Wherefore (by Law II), the effects of those collisions will be equal in both cases; and therefore the mutual motions of the bodies among themselves in the one case will remain equal to the mutual motions of the bodies among themselves in the other. A clear proof of which we have from the experiment of a ship; where all motions happen after the same manner, whether the ship is at rest, or is carried uniformly forwards in a right line.

COROLLARY VI. IF BODIES, ANY HOW MOVED AMONG THEMSELVES, ARE URGED IN THE DIRECTION OF PARALLEL LINES BY EQUAL ACCELERATIVE FORCES, THEY WILL ALL CONTINUE TO MOVE AMONG THEMSELVES, AFTER THE SAME MANNER AS IF THEY HAD BEEN URGED BY NO SUCH FORCES.

For these forces acting equally (with respect to the quantities of the bodies to be moved), and in the direction of parallel lines, will (by Law II) move all the bodies equally (as to velocity), and therefore will never produce any change in the positions or motions of the bodies among themselves.

SCHOLIUM.

Hitherto I have laid down such principles as have been received by mathematicians, and are confirmed by abundance of experiments. By the first two Laws and the first two Corollaries, Galileo discovered that the descent of bodies observed the duplicate ratio of the time, and that the motion of projectiles was in the curve of a parabola; experience agreeing with both, unless so far as these motions are a little retarded by the resistance of the air. When a body is falling, the uniform force of its gravity acting equally, impresses, in equal particles of time, equal forces upon that body, and therefore generates equal velocities; and in the whole time impresses a whole force, and generates a whole velocity proportional to the time. And the spaces described in proportional times are as the velocities and the times conjunctly; that is, in a duplicate ratio of the times. And when a body is thrown upwards, its uniform gravity impresses forces and takes off velocities proportional to the times; and the times of ascending to the greatest heights are as the velocities to be taken off, and those heights are as the velocities and the times conjunctly, or in the duplicate ratio of the velocities. And if a body be projected in any direction, the motion arising from its projection is compounded with the motion arising from its gravity. As if the body A by its motion of projection alone could describe in a given time the right line AB, and with its motion of falling alone could describe in the same time the altitude AC; complete the paralellogram ABDC, and the body by that compounded motion will at the end of the time be found in the place D; and the curve line AED, which that body describes, will be a parabola, to which the right line AB will be a tangent in A; and whose ordinate BD will be as the square of the line AB. On the same Laws and Corollaries depend those things which have been demonstrated concerning the times of the vibration of pendulums, and are confirmed by the daily experiments of pendulum clocks. By the same, together with the third Law, Sir Christ. Wren, Dr. Wallis, and Mr. Huygens, the greatest geometers of our times, did severally determine the rules of the congress and reflection of hard bodies, and much about the same time communicated their discoveries to the Royal Society, exactly agreeing among themselves as to those rules. Dr. Wallis, indeed, was something more early in the publication; then followed Sir Christopher Wren, and, lastly, Mr. Huygens. But Sir Christopher Wren confirmed the truth of the thing before the Royal Society by the experiment of pendulums, which Mr. Mariotte soon after thought fit to explain in a treatise entirely upon that subject. But to bring this experiment to an accurate agreement with the theory, we are to have a due regard as well to the resistance of the air as to the elastic force of the concurring bodies. Let the spherical bodies A, B be suspended by the parallel and equal strings AC, BD, from the centers C, D. About these centers, with those intervals, describe the semicircles EAF, GBH, bisected by the radii CA, DB. Bring the body A to any point R of the LAP, and (withdrawing the body B) let it go from thence, and after one oscillation suppose it to return to the point V: then RV will be the retardation arising from the resistance of the air. Of this RV let ST be a fourth part, situated in the middle, to wit, so as RS and TV may be equal, and RS may be to ST as 3 to 2 then will ST represent very nearly the retardation during the descent from S to A. Restore the body B to its place: and, supposing the body A to be let fall from the point S, the velocity thereof in the place of reflection A, without sensible error, will be the same as if it had descended in vacuo from the point T. Upon

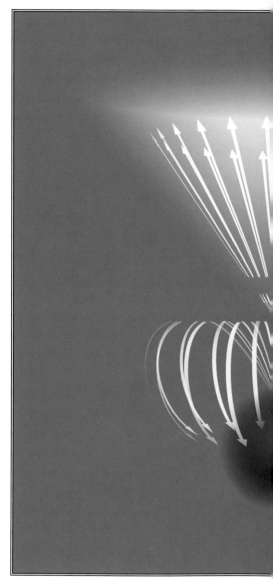

which account this velocity may be represented by the chord of the arc TA. For it is a proposition well known to geometers, that the velocity of a pendulous body in the lowest point is as the chord of the arc which it has described in its descent. After reflection, suppose the body A comes to the place s, and the body B to the place k. Withdraw the body B, and find the place v, from which if the body A, being let go, should after one oscillation return to the place r, st may be a fourth part of rv, so placed

in the middle thereof as to leave rs equal to tv, and let the chord of the arc tA represent the velocity which the body A had in the place A immediately after reflection. For t will be the true and correct place to which the body A should have ascended, if the resistance of the air had been taken off. In the same way we are to correct the place k to which the body B ascends, by finding the place l to which it should have ascended in vacuo. And thus everything may be subjected to experiment, in the same manner as if we were really placed in vacuo. These things being done, we are to take the product (if I may so say) of the body A, by the chord of the arc TA (which represents its velocity), that we may have its motion in the place A immediately before reflection; and then by the chord of the arc tA, that we may have its motion in the place A immediately after reflection. And so we are to take the product of the body B by the chord of the arc Bl that we may have the motion of the same immediately after reflection. And in like manner, when two bodies are let go together from different places, we are to find the motion of each, as well before as after reflection; and then we may compare the motions between themselves, and collect the effects of the reflection. Thus trying the thing with pendulums of ten feet, in unequal as well as

Newtonian theory of gravity can even contribute to our understanding of what happens when a star collapses under its own gravitational field.

In a standard situation, a star balances the nuclear and the gravitational forces. Light escapes the surface of the star.

As the star loses its nuclear gravity, it begins to act on the escaping light.

As the star collapses, the light is drawn back to the surface.

Finally, the gravitational field of the collapsed star is too powerful for the light to escape, creating what we know as a black hole.

All this is implied in Newton's original theories, although it was not fully proposed until long after his death.

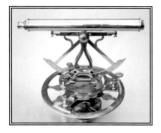

18th-century English telescope and compass.

equal bodies, and making the bodies to concur after a descent through large spaces, as of 8, 12, or 16 feet, I found always, without an error of 3 inches, that when the bodies concurred together directly, equal changes towards the contrary parts were produced in their motions, and, of consequence, that the action and reaction were always equal. As if the body A impinged upon the body B at rest with 9 parts of motion, and losing 7, proceeded after reflection with 2, the body B was carried backwards with those 7 parts. If the bodies concurred with contrary motions, A with twelve parts of motion, and B with six, then if A receded with 2, B receded with 8; to wit, with a deduction of 14 parts of motion on each side. For from the motion of A subducting twelve parts, nothing will remain; but subducting 2 parts more, a motion will be generated of 2 parts towards the contrary way; and so, from the motion of the body B of 6 parts, subducting 14 parts, a motion is generated of 8 parts towards the contrary way. But if the bodies were made both to move towards the same way, A, the swifter, with 14 parts of motion, B, the slower, with 5, and after reflection A went on with 5, B likewise went on with 14 parts; 9 parts being transferred from A to B. And so in other cases. By the congress and collision of bodies, the quantity of motion, collected from the sum of the motions directed towards the same way, or from the difference of those that were directed towards contrary ways, was never changed. For the error of an inch or two in measures may be easily ascribed to the difficulty of executing everything with accuracy. It was not easy to let go the two pendulums so exactly together that the bodies should impinge one upon the other in the lowermost place AB; nor to mark the places s, and k, to which the bodies ascended after congress. Nay, and some errors, too, might have happened from the unequal density of the parts of the pendulous bodies themselves, and from the irregularity of the texture proceeding from other causes.

But to prevent an objection that may perhaps be alleged against the rule, for the proof of which this experiment was made, as if this rule did suppose that the bodies were either absolutely hard, or at least perfectly elastic (whereas no such bodies are to be found in nature), I must add, that the experiments we have been describing, by no means depending upon that quality of hardness, do succeed as well in soft as in hard bodies. For if the rule is to be tried in bodies not perfectly hard, we are only to diminish the reflection in such a certain proportion as the quantity of the elastic force requires. By the theory of Wren and Huygens, bodies absolutely hard return one from another with the same velocity with which they meet. But this may be affirmed with more certainty of bodies perfectly elastic. In bodies imperfectly elastic the velocity of the return is to be diminished together with the elastic force; because that force (except when the parts of bodies are bruised by their congress, or suffer some such extension as happens under the strokes of a hammer) is (as far as I can perceive) certain and determined, and makes the bodies to return one from the other with a relative velocity, which is in a given ratio to that relative velocity with which they met. This I tried in balls of wool, made up tightly, and strongly compressed. For, first, by letting go the pendulous bodies, and measuring their reflection, I determined the quantity of their elastic force; and then, according to this force, estimated the reflections that ought to happen in other cases of congress. And with this computation other experiments made afterwards did accordingly agree; the balls always receding one from the other with a relative velocity, which was to the relative velocity with which they met as about 5 to 9. Balls of steel returned with almost the same velocity: those of cork with a velocity something less; but in balls of glass the proportion was as about 15 to 16. And thus the third Law, so far as it regards percussions and reflections, is proved by a theory exactly agreeing with experience.

In attractions, I briefly demonstrate the thing after this manner. Suppose an obstacle is interposed to hinder the congress of any two bodies A, B, mutually attracting one the other: then if either body, as A, is more attracted towards the other body B, than that other body B is towards the first body A, the obstacle will be more strongly urged by the pressure of the body A than by the pressure of the body B, and therefore will not remain in equilibrio: but the stronger pressure will prevail, and will make the system of the two bodies, together with the obstacle, to move directly towards the parts on which B lies; and in free spaces, to go forward in infinitum with a motion perpetually accelerated; which is absurd and contrary to the first Law. For, by the first Law, the system ought to persevere in its state of rest, or of moving uniformly forward in a right line; and therefore the bodies must equally press the obstacle, and be equally attracted one by the other. I made the experiment on the loadstone and iron. If these, placed apart in proper vessels, are made to float by one another in standing water, neither of them will propel the other; but, by being equally attracted, they will sustain each other's pressure, and rest at last in an equilibrium.

So the gravitation betwixt the earth and its parts is mutual. Let the earth FI be cut by any plane EG into two parts EGF and EGI, and their weights one towards the other will be mutually equal. For if by another plane HK, parallel to the former EG, the greater part EGI is cut into two parts EGKH and HKI, whereof HKI is equal to the part EFG, first cut off, it is evident that the middle part EGKH, will have no propension by its proper weight towards either side, but will hang as it were, and rest in an equilibrium betwixt both. But the one extreme part HKI will with its whole weight bear upon and press the middle part towards the other extreme part EGF; and therefore the force with which EGI, the sum of the parts HKI and EGKH, tends towards the third part EGF, is equal to the weight of the part HKI, that is, to the weight of the third part EGF. And therefore the weights of the two parts EGI and EGF, one towards the other, are equal, as I was to prove. And indeed if those weights were not equal, the whole Earth floating in the nonresisting ether would give way to the greater weight, and, retiring from it, would be carried off *in infinitum*.

And as those bodies are equipollent in the congress and reflection, whose velocities are reciprocally as their innate forces, so in the use of mechanic instruments those agents are equipollent, and mutually sustain each the contrary pressure of the other, whose velocities, estimated according to the determination of the forces, are reciprocally as the forces.

So those weights are of equal force to move the arms of a balance; which during the play of the balance are reciprocally as their velocities

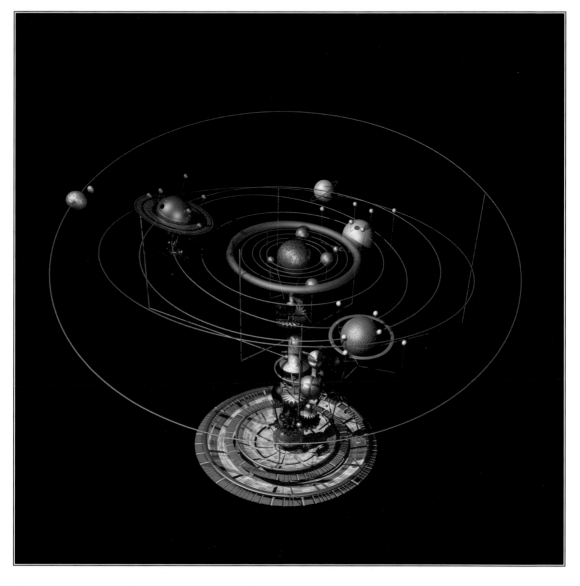

Newtonian-style orrery with the later discovered asteroid belt.

upwards and downwards; that is, if the ascent or descent is direct, those weights are of equal force, which are reciprocally as the distances of the points at which they are suspended from the axis of the balance; but if they are turned aside by the interposition of oblique planes, or other obstacles, and made to ascend or descend obliquely, these bodies will be equipollent, which are reciprocally as the heights of their ascent and descent taken according to the perpendicular; and that on account of the determination of gravity downwards.

And in like manner in the pulley, or in a combination of pullies, the force of a hand drawing the rope directly, which is to the weight, whether ascending directly or obliquely, as the velocity of the perpendicular ascent of the weight to the velocity of the hand that draws the rope, will sustain the weight.

In clocks and such like instruments, made up from a combination of wheels, the contrary forces that promote and impede the motion of the wheels, if they are reciprocally as the velocities of the parts of the wheel on which they are impressed, will mutually sustain the one the other.

The force of the screw to press a body is to the force of the hand that turns the handles by which it is moved as the circular velocity of the handle in that part where it is impelled by the hand is to the progressive velocity of the screw towards the pressed body.

The forces by which the wedge presses or drives the two parts of the wood it cleaves are to the force of the mallet upon the wedge as the progress of the wedge in the direction of the force impressed upon it by the mallet is to the velocity with which the parts of the wood yield to the wedge, in the direction of lines perpendicular to the sides of the wedge. And the like account is to be given of all machines.

The power and use of machines consist only in this, that by diminishing the velocity we may augment the force, and the contrary: from whence in all sorts of proper machines, we have the solution of this problem; To move a given weight with a given power, or with a given force to overcome any other given resistance. For if machines are so contrived that the velocities of the agent and resistant are reciprocally as their forces, the agent will just sustain the resistant, but with a greater disparity of velocity will overcome it. So that if the disparity of velocities is so great as to overcome all that resistance which commonly arises either from the attrition of contiguous bodies as they slide by one another, or from the cohesion of continuous bodies that are to be separated, or from the weights of bodies to be raised, the excess of the force remaining, after all those resistances are overcome, will produce an acceleration of motion proportional thereto, as well in the parts of the machine as in the resisting body. But to treat of mechanics is not my present business. I was only willing to show by those examples the great extent and certainty of the third Law of motion. For if we estimate the action of the agent from its force and velocity conjunctly, and likewise the reaction of the impediment conjunctly from the velocities of its several parts, and from the forces of resistance arising from the attrition, cohesion, weight, and acceleration of those parts. the action and reaction in the use of all sorts of machines will be found always equal to one another. And so far as the action is propagated by the intervening instruments, and at last impressed upon the resisting body, the ultimate determination of the action will be always contrary to the determination of the reaction.

———

The spacecraft Cassini's interplanetary trajectory. A spacecraft requires complex mathematics to calculate trajectories, orbits, and slingshot effects. All these are based squarely on Newton's theoretical models, which are over three hundred years old. The complexities of calculated orbits and final launching of the Titan probe remain a remarkable testimony to Newton's contribution to science.

BOOK III.

———

RULES OF REASONING IN PHILOSOPHY.

RULE I. WE ARE TO ADMIT NO MORE CAUSES OF NATURAL THINGS THAN SUCH AS ARE BOTH TRUE AND SUFFICIENT TO EXPLAIN THEIR APPEARANCES.

To this purpose the philosophers say that Nature does nothing in vain, and more is in vain when less will serve; for Nature is pleased with simplicity, and affects not the pomp of superfluous causes.

RULE II. THEREFORE TO THE SAME NATURAL EFFECTS WE MUST, AS FAR AS POSSIBLE, ASSIGN THE SAME CAUSES.

As to respiration in a man and in a beast; the descent of stones in Europe and in America; the light of our culinary fire and of the Sun; the reflection of light in the Earth, and in the planets.

RULE III. THE QUALITIES OF BODIES, WHICH ADMIT NEITHER INTENSION NOR REMISSION OF DEGREES, AND WHICH ARE FOUND TO BELONG TO ALL BODIES WITHIN THE REACH OF OUR EXPERIMENTS, ARE TO BE ESTEEMED THE UNIVERSAL QUALITIES OF ALL BODIES WHATSOEVER.

For since the qualities of bodies are only known to us by experiments, we are to hold for universal all such as universally agree with experiments; and such as are not liable to diminution can never be quite taken away. We are certainly not to relinquish the evidence of experiments for the sake of dreams and vain fictions of our own devising; nor are we to recede from the analogy of Nature, which uses to be simple, and always consonant to itself. We no other way know the extension of bodies than by our senses, nor do these reach it in all bodies; but because we perceive extension in all that are sensible, therefore we ascribe it universally to all others also. That abundance of bodies are hard, we learn by experience; and because the hardness of the whole arises from the hardness of the parts, we therefore justly infer the hardness of the undivided particles not only of the bodies we feel but of all others. That all bodies are impenetrable, we gather not from reason, but from sensation. The bodies which we handle we find impenetrable, and thence conclude impenetrability to be an universal property of all bodies whatsoever. That all bodies are moveable, and endowed with certain powers (which we call the *vires inertiae*) of persevering in their motion, or in their rest, we only infer from the like properties observed in the bodies which we have seen. The extension, hardness, impenetrability, mobility, and *vis inertia* of the whole, result from the extension, hardness, impenetrability, mobility, and *vires inertia* of the parts; and thence we conclude the least particles of all bodies to be also all extended, and hard and impenetrable, and moveable, and endowed with their proper *vires inertiae*. And this is the foundation of all philosophy. Moreover, that the divided but contiguous particles of bodies may be separated from one another, is matter of observation; and, in the particles that remain undivided, our minds are able to distinguish yet lesser parts, as is mathematically demonstrated. But whether the parts so distinguished, and not yet divided, may, by the powers of Nature, be actually divided and separated from one another, we cannot certainly determine. Yet, had we the proof of but one experiment that any undivided particle, in breaking a hard and solid body, suffered a division, we might by virtue of this rule conclude that the undivided as well as the divided particles may be divided and actually separated to infinity.

Lastly, if it universally appears, by experiments and astronomical observations, that all bodies about the Earth gravitate towards the Earth, and that in proportion to the quantity of matter which they severally contain, that the Moon likewise, according to the quantity of its matter, gravitates towards the Earth; that, on the other hand, our sea gravitates towards the Moon; and all the planets mutually one towards another; and the comets in like manner towards the Sun; we must in consequence of this rule, universally allow that all bodies whatsoever are endowed with a principle of mutual gravitation. For the argument from the appearances concludes with more force for the universal gravitation of all bodies than for their impenetrability; of which, among those in the celestial regions, we have no experiments, nor any manner of observation. Not that I affirm gravity to be essential to bodies: by their *vis insita* I mean nothing but their *vis inertiae*. This is immutable. Their gravity is diminished as they recede from the earth.

RULE IV. IN EXPERIMENTAL PHILOSOPHY WE ARE TO LOOK UPON PROPOSITIONS COLLECTED BY GENERAL INDUCTION FROM PHENOMENA AS ACCURATELY OR VERY NEARLY TRUE, NOTWITHSTANDING ANY CONTRARY HYPOTHESES THAT MAY BE IMAGINED, TILL SUCH TIME AS OTHER PHENOMENA OCCUR, BY WHICH THEY MAY EITHER BE MADE MORE ACCURATE, OR LIABLE TO EXCEPTIONS.

This rule we must follow, that the argument of induction may not be evaded by hypotheses.

———

OF THE MOTION OF THE MOON'S NODES.

PROPOSITION I. THE MEAN MOTION OF THE SUN FROM THE NODE IS DEFINED BY A GEOMETRIC MEAN PROPORTIONAL BETWEEN THE MEAN MOTION OF THE SUN AND THAT MEAN MOTION WITH WHICH THE SUN RECEDES WITH THE GREATEST SWIFTNESS FROM THE NODE IN THE QUADRATURES.

"Let T be the Earth's place, N*n* the line of the Moon's nodes at any given time, KTM a perpendicular thereto, TA a right line revolving about the center with the same angular velocity with which the Sun and the node recede from one another, in such sort that the angle between the quiescent right line N*n* and the revolving line TA may be always equal to the distance of the places of the Sun and node. Now if any right line TK be divided into parts TS and SK, and those parts be taken as the mean horary motion of the Sun to the mean horary motion of the node in the quadratures, and there be taken the right line TH, a mean proportional between the part TS and the whole TK, this right line will be proportional to the Sun's mean motion from the node.

"For let there be described the circle NK*n*M from the center T and with the radius TK, and about the same center, with the semi-axis TH

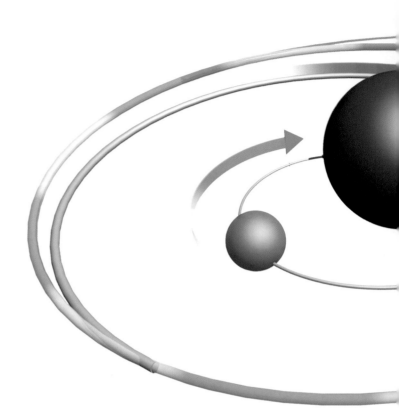

and TN, let there be described an ellipsis NH*n*L; and in the time in which the Sun recedes from the node through the arc N*a*, if there be drawn the right line T*ba*, the area of the sector NT*a* will be the exponent of the sum of the motions of the Sun and node in the same time. Let, therefore, the extremely small arc *a*A be that which the right line T*ba*, revolving according to the aforesaid law, will uniformly describe in a given particle of time, and the extremely small sector TA*a* will be as the sum of the velocities with which the sun and node are carried two different ways in that time. Now the Sun's velocity is almost uniform, its inequality being so small as scarcely to produce the least inequality in the mean motion of the nodes. The other part of this sum, namely, the mean quantity of the velocity of the node, is increased in the recess from the syzygies in a duplicate ratio of the sine of its distance from the Sun (by Cor. Prop. XXXI, of this Book), and, being greatest in its quadratures with the Sun

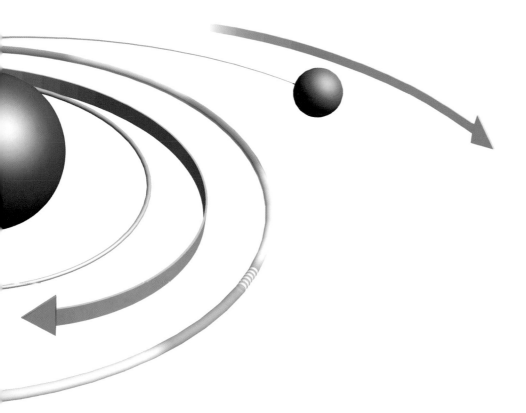

in K, is in the same ratio to the Sun's velocity as SK to TS, that is, as (the difference of the squares of TK and TH, or) the rectangle KHM to TH². But the ellipsis NBH divides the sector ATa, the exponent of the sum of these two velocities, into two parts ABba and BTb, proportional to the velocities. For produce BT to the circle in β, and from the point B let fall upon the greater axis the perpendicular BG, which being produced both ways may meet the circle in the points F and ƒ; and because the space ABba is to the sector TBb as the rectangle Abβ to BT² (that rectangle being equal to the difference of the squares of TA and TB, because the right line Aβ is equally cut in T, and unequally in B), therefore when the space ABba is the greatest of all in K, this ratio will be the same as the ratio of the rectangle KHM to HT². But the greatest mean velocity of the node was shown above to be in that very ratio to the velocity of the Sun; and therefore in the quadratures the sector ATa is divided into parts

If the force of gravity was less, or increased more rapidly with distance than Newton's theory predicts, the orbits of the planets around the sun would not be stable ellipses. They would either fly away from the sun, or spiral in.

proportional to the velocities. And because the rectangle KHM is to HT^2 as FBf to BG^2, and the rectangle ABβ is equal to the rectangle FBf, therefore the little area ABba, where it is greatest, is to the remaining sector TBb as the rectangle AB to BG^2. But the ratio of these little areas always was as the rectangle AB(to BT^2; and therefore the little area ABba in the place A is less than its correspondent little area in the quadratures in the duplicate ratio of BG to BT, that is, in the duplicate ratio of the sine of the Sun's distance from the node. And therefore the sum of all the little areas ABba, to wit, the space ABN, will be as the motion of the node in the time in which the Sun hath been going over the arc NA since he left the node; and the remaining space, namely, the elliptic sector NTB, will be as the Sun's mean motion in the same time. And because the mean annual motion of the node is that motion which it performs in the time that the Sun completes one period of its course, the mean motion of the mode from the Sun will be to the mean motion of the Sun itself as the area of the circle to the area of the ellipsis; that is, as the right line TK to the right line TH, which is a mean proportional between TK and TS; or, which comes to the same as the mean proportional TH to the right line TS.

———

PROPOSITION XXXVI. PROBLEM XVII.

TO FIND THE FORCE OF THE SUN TO MOVE THE SEA.

The Sun's force ML or PT to disturb the motions of the Moon, was (by Prop. XXV.) in the Moon's quadratures, to the force of gravity with us, as 1 to 638092,6; and the force TM − LM or 2PK in the Moon's syzygies is double that quantity. But, descending to the surface of the Earth, these forces are diminished in proportion of the distances from the Center of the Earth, that is, in the proportion of 601/2 to 1; and therefore the former force on the Earth's surface is to the force of gravity as 1 to 38604600; and by this force the sea is depressed in such places as are 90 degrees distant from the Sun, But by the other force, which is twice as great, the sea is raised not only in the places directly under the Sun, but in those also which are directly opposed to it; and the sum of these forces is to the force of gravity as 1 to 12868200. And because the same force excites the same motion, whether it depresses the waters in those places which are 90 degrees distant from the sun, or raises them in the places which are directly under and directly opposed to the sun, the aforesaid sum will be the total force of the Sun to disturb the sea, and will have the same effect as if the whole was employed in raising the sea in the places directly under and directly opposed to the Sun, and did not act at all in the places which are 90 degrees removed from the Sun.

And this is the force of the Sun to disturb the sea in any given place, where the Sun is at the same time both vertical, and in its mean distance from the Earth. In other positions of the Sun, its force to raise the sea is as the versed sine of double its altitude above the horizon of the place directly, and the cube of the distance from the earth reciprocally.

COR. Since the centrifugal force of the parts of the Earth, arising from the Earth's diurnal motion, which is to the force of gravity as 1 to 289, raises the waters under the equator to a height exceeding that under the poles by 85472 *Paris* feet, as above, in Prop. XIX., the force of the sun, which we have now showed to be to the force of gravity as 1 to 12868200, and therefore is to that centrifugal force as 289 to 12868200, or as 1 to 44527, will be able to raise the waters in the places directly under and directly opposed to the Sun to a height exceeding that in the places which are 90 degrees removed from the Sun only by one *Paris* foot and 1131/30 inches; for this measure is to the measure of 85472 feet as 1 to 44527.

―――――

PROPOSITION XXXVIII. PROBLEM XIX.

TO FIND THE FIGURE OF THE MOONS' BODY.

If the Moon's body were fluid like our sea, the force of the Earth to raise that fluid in the nearest and remotest parts would be to the force of the Moon by which our sea is raised in the places under and opposite to the Moon as the accelerative gravity of the Moon towards the Earth to the accelerative gravity of the Earth towards the Moon, and the diameter of the Moon to the diameter of the Earth conjunctly; that is, as 39,788 to 1, and 100 to 365 conjunctly, or as 1081 to 100. Wherefore, since our sea, by the force of the Moon, is raised to 83/5 feet, the lunar fluid would be raised by the force of the Earth to 93 feet; and upon this account the figure of the Moon would be a spheroid, whose greatest

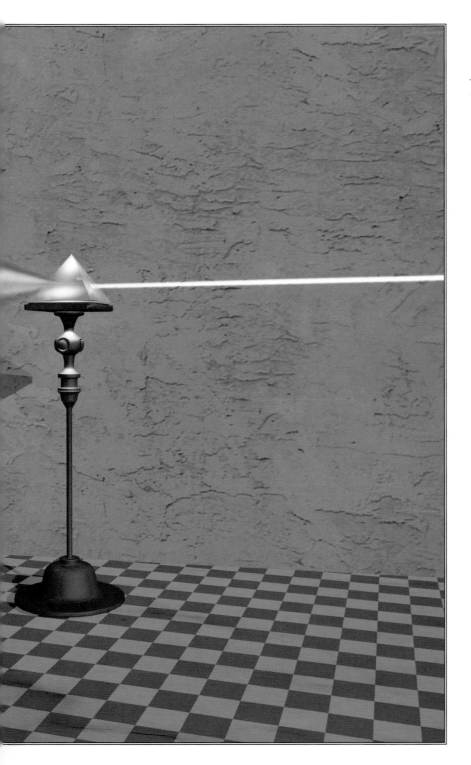

One of Newton's greatest discoveries was in the science of optics. He found that if light from the Sun passes through a prism, it breaks up into its component colors (spectrum), the colors of the rainbow.

OPPOSITE PAGE

Cassini space craft launching probe which parachutes down on Titan, one of Saturn's moons.

diameter produced would pass through the center of the Earth, and exceed the diameters perpendicular thereto by 186 feet. Such a figure, therefore, the Moon affects, and must have put on from the beginning.

Q.E.I.

COR. Hence it is that the same face of the Moon always respects the earth; nor can the body of the Moon possibly rest in any other position, but would return always by a libratory motion to this situation; but those librations, however, must be exceedingly slow, because of the weakness of the forces which excite them; so that the face of the Moon, which should be always obverted to the earth, may, for the reason assigned in Prop. XVII. be turned towards the other focus of the Moon's orbit, without being immediately drawn back, and converted again towards the Earth.

———

END OF THE MATHEMATICAL PRINCIPLES

Albert Einstein (1879-1955)

HIS LIFE AND WORK

Genius isn't always immediately recognized. Although Albert Einstein would become the greatest theoretical physicist who ever lived, when he was in grade school in Germany his headmaster told his father, "He'll never make a success of anything." When Einstein was in his mid-twenties, he couldn't find a decent teaching job even though he had graduated from the Federal Polytechnic School in Zurich as a teacher of mathematics and physics. So he gave up hope of obtaining a university position and applied for temporary work in Bern. With the help of a classmate's father, Einstein managed to secure a civil-service post as an examiner in the Swiss patent office. He worked six days a week, earning $600 a year. That's how he supported himself while working toward his doctorate in physics at the University of Zurich.

In 1903, Einstein married his Serbian sweetheart, Mileva Maric, and the couple moved into a one-bedroom flat in Bern. Two years later, she bore him a son, Hans Albert. The period surrounding Hans's birth was probably the happiest time in Einstein's life. Neighbors later recalled seeing the young father absentmindedly pushing a baby carriage down the city streets. From time to time, Einstein would reach into the carriage and remove a pad of paper on which to jot down notes to himself. It seems likely that the notepad in the baby's stroller contained some of the formulas and equations that led to the theory of relativity and the development of the atomic bomb.

During these early years at the patent office, Einstein spent most of his spare time studying theoretical physics. He composed a series of four

The young Einstein.

seminal scientific papers, which set forth some of the most momentous ideas in the long history of the quest to comprehend the universe. Space and time would never be looked at the same way again. Einstein's work won him the Nobel Prize in Physics in 1921, as well as much popular acclaim.

As Einstein pondered the workings of the universe, he received flashes of understanding that were too deep for words. "These thoughts did not come in any verbal formulation," Einstein was once quoted as saying. "I rarely think in words at all. A thought comes, and I may try to express it in words afterward."

Einstein eventually settled in the United States, where he publicly championed such causes as Zionism and nuclear disarmament. But he maintained his passion for physics. Right up until his death in 1955, Einstein kept seeking a unified field theory that would link the phenomena of gravitation and electromagnetism in one set of equations. It is a tribute to Einstein's vision that physicists today continue to seek a grand unification of physical theory. Einstein revolutionized scientific thinking in the twentieth century and beyond.

Albert Einstein was born at Ulm, in the former German state of Wüettemberg, on March 14, 1879, and grew up in Munich. He was the only son of Hermann Einstein and Pauline Koch. His father and uncle owned an electrotechnical plant. The family considered Albert a slow learner because he had difficulty with language. (It is now thought that he may have been dyslexic.) Legend has it that when Hermann asked the headmaster of his son's school about the best profession for Albert, the man replied, "It doesn't matter. He'll never make a success of anything."

Einstein did not do well in school. He didn't like the regimentation, and he suffered from being one of the few Jewish children in a Catholic school. This experience as an outsider was one that would repeat itself many times in his life.

One of Einstein's early loves was science. He remembered his father's showing him a pocket compass when he was around five years old, and marveling that the needle always pointed north, even if the case was spun. In that moment, Einstein recalled, he "felt something deeply hidden had to be behind things." Another of his early loves was music. Around the age of six, Einstein began studying the violin. It did not come naturally to him; but when after several years he recognized the mathematical structure of music, the violin became a lifelong passion— although his talent was never a match for his enthusiasm.

When Einstein was ten, his family enrolled him in the Luitpold Gymnasium, which is where, according to scholars, he developed a suspicion of authority. This trait served Einstein well later in life as a scientist. His habit of skepticism made it easy for him to question many long-standing scientific assumptions.

In 1895, Einstein attempted to skip high school by passing an entrance examination to the Federal Polytechnic School in Zurich, where he hoped to pursue a degree in electrical engineering. This is what he wrote about his ambitions at the time:

If I were to have the good fortune to pass my examinations, I would go to Zurich. I would stay there for four years in order to study mathematics and physics. I imagine myself becoming a teacher in those branches of the natural sciences, choosing the theoretical part of them. Here are the reasons which lead me to this plan. Above all, it is my disposition for abstract and mathematical thought, and my lack of imagination and practical ability.

Einstein failed the arts portion of the exam and so was denied admission to the polytechnic. His family instead sent him to secondary school at Aarau, in Switzerland, hoping that it would earn him a second chance to enter the Zurich school. It did, and Einstein graduated from the polytechnic in 1900. At about that time he fell in love with Mileva Maric, and in 1901, she gave birth out of wedlock to their first child, a daughter named Lieserl. Very little is known for certain about Lieserl, but it appears that she either was born with a crippling condition or fell very ill as an infant, then was put up for adoption, and died at about two years of age. Einstein and Maric married in 1903.

The year Hans was born, 1905, was a miracle year for Einstein. Somehow he managed to handle the demands of fatherhood and a full-time job and still publish four epochal scientific papers, all without benefit of the resources that an academic appointment might have provided.

In the spring of that year, Einstein submitted three papers to the German periodical *Annals of Physics* (*Annalen der Physik*). The three appeared together in the journal's volume 17. Einstein characterized the first paper, on the light quantum, as "very revolutionary." In it, he examined the phenomenon of the quantum (the fundamental unit of energy) discovered by the German physicist Max Planck. Einstein explained the photoelectric effect, which holds that for each electron emitted, a specific amount of energy is released. This is the quantum effect that states that energy is emitted in fixed amounts that can be expressed only as whole integers. This theory formed the basis for a great deal of quantum

Einstein with his first wife, Mileva and their son, Hans Albert, 1906.

mechanics. Einstein suggested that light be considered a collection of independent particles of energy, but remarkably, he offered no experimental data. He simply argued hypothetically for the existence of these "light quantum" for aesthetic reasons.

Initially, physicists were hesitant to endorse Einstein's theory. It was too great a departure from scientifically accepted ideas of the time, and far beyond anything Planck had discovered. It was this first paper, titled "On a Heuristic View concerning the Production and Transformation of Light"—not his work on relativity—that won Einstein the Nobel Prize in Physics in 1921.

In his second paper, "On a New Determination of Molecular Dimensions"—which Einstein wrote as his doctoral dissertation—and his third, "On the Movement of Small Particles Suspended in Stationary Liquids Required by the Molecular-Kinetic Theory of Heat," Einstein proposed a method to determine the size and motion of atoms. He also explained Brownian motion, a phenomenon described by the British botanist Robert Brown after studying the erratic movement of pollen suspended in fluid. Einstein asserted that this movement was caused by impacts between atoms and molecules. At the time, the very existence of atoms was still a subject of scientific debate, so there could be no underestimating the importance of these two papers. Einstein had confirmed the atomic theory of matter.

In the last of his 1905 papers, entitled "On the Electrodynamics of Moving Bodies," Einstein presented what became known as the special theory of relativity. The paper reads more like an essay than a scientific communication. Entirely theoretical, it contains no notes or bibliographic citations. Einstein wrote this 9,000-word treatise in just five weeks, yet historians of science consider it every bit as comprehensive and revolutionary as Isaac Newton's *Principia*.

What Newton had done for our understanding of gravity, Einstein had done for our view of time and space, managing in the process to overthrow the Newtonian conception of time. Newton had declared that "absolute, true, and mathematical time, of itself and from its own nature, flows equably without relation to anything external." Einstein held that all observers should measure the same speed for light, regardless of how fast they themselves are moving. Einstein also asserted that the mass of an object is not unchangeable but rather increases with the object's velocity. Experiments later proved that a small particle of matter, when accelerated to 86 percent of the speed of light, has twice as much mass as it does at rest.

Another consequence of relativity is that the relation between energy and mass may be expressed mathematically, which Einstein did in the famous equation $E=mc2$. This expression—that energy is equivalent to mass times the square of the speed of light—led physicists to understand that even miniscule amounts of matter have the potential to yield enormous amounts of energy. Completely converting to energy just a part of the mass of a few atoms would, then, result in a colossal explosion. Thus did Einstein's modest-looking equation lead scientists to consider the consequences of splitting the atom (nuclear fission) and, at the urging of governments, to develop the atomic bomb. In 1909, Einstein was appointed professor of theoretical physics at the University of Zurich, and three years later he fulfilled his ambition to return to the Federal Polytechnic School as a full professor. Other prestigious academic appointments and directorships followed. Throughout, he continued to work on his theory of gravity as well as his general theory of relativity. But as his professional status continued to rise, his marriage and health began to deteriorate. He and Mileva began divorce proceedings in 1914, the same year he accepted a professorship at the University of Berlin. When he later fell ill, his cousin Elsa nursed him back to health, and around 1919 they were married.

Where the special theory of relativity radically altered concepts of time and mass, the general theory of relativity changed our concept of space. Newton had written that "absolute space, in its own nature, without relation to anything external, remains always similar and immovable." Newtonian space is Euclidean, infinite, and unbounded. Its geometric structure is completely independent of the physical matter occupying it. In it, all bodies gravitate toward one another without having any effect on the structure of space. In stark contrast, Einstein's general theory of relativity asserts that not only does a body's gravitational mass act on other bodies, it also influences the structure of space. If a body is massive enough, it induces space to curve around it. In such a region, light appears to bend.

In 1919, Sir Arthur Eddington sought evidence to test the general theory. Eddington organized two expeditions, one to Brazil and the other to West Africa, to observe the light from stars as it passed near a massive body—the Sun—during a total solar eclipse on May 29. Under normal circumstances such observations would be impossible, as the weak light from distant stars would be blotted out by daylight, but during the eclipse such light would briefly be visible.

ABOVE

Einstein and his second wife, Elsa with Charlie Chaplin, 1931.

OPPOSITE PAGE

Einstein near the time when he won the Nobel prize.

Einstein teaching at Princeton in 1932.

In September, Einstein received a telegram from Hendrik Lorentz, a fellow physicist and close friend. It read: "Eddington found star displacement at rim of Sun, preliminary measurements between nine-tenths of a second and twice that value." Eddington's data were in keeping with the displacement predicted by the special relativity theory. His photographs from Brazil seemed to show the light from known stars in a different position in the sky during the eclipse than they were at nighttime, when their light did not pass near the Sun. The theory of general relativity had been confirmed, forever changing the course of physics. Years later, when a student of Einstein's asked how he would have reacted had the theory been disproved, Einstein replied, "Then I would have felt sorry for the dear Lord. The theory is correct."

Confirmation of general relativity made Einstein world-famous. In 1921, he was elected a member of the British Royal Society. Honorary degrees and awards greeted him at every city he visited. In 1927, he began developing the foundation of quantum mechanics with the Danish physicist Niels Bohr, even as he continued to pursue his dream of a unified field theory. His travels in the United States led to his appointment in 1932 as a professor of mathematics and theoretical physics at the Institute for Advanced Study in Princeton, New Jersey.

A year later, he settled permanently in Princeton after the ruling Nazi party in Germany began a campaign against "Jewish science." Einstein's property was confiscated, and he was deprived of German citizenship and positions in German universities. Until then, Einstein had considered himself a pacifist. But when Hitler turned Germany into a military power in Europe, Einstein came to believe that the use of force against Germany was justified. In 1939, at the dawn of World War II, Einstein became concerned that the Germans might be developing the capability to build an atomic bomb—a weapon made possible by his own research and for which he therefore felt a responsibility. He sent a letter to President Franklin D. Roosevelt warning of such a possibility and urging that the United States undertake nuclear research. The letter, composed by his friend and fellow scientist Leo Szilard, became the impetus for the formation of the Manhattan Project, which produced the world's first atomic weapons. In 1944, Einstein put a handwritten copy of his 1905 paper on special relativity up for auction and donated the proceeds—six million dollars—to the Allied war effort.

After the war, Einstein continued to involve himself with causes and issues that concerned him. In November 1952, having shown strong support for Zionism for many years, he was asked to accept the presidency of Israel. He respectfully declined, saying that he was not suited for the position. In April 1955, only one week before his death, Einstein composed a letter to the philosopher Bertrand Russell in which he agreed to sign his name to a manifesto urging all nations to abandon nuclear weapons.

Einstein died of heart failure on April 18, 1955. Throughout his life, he had sought to understand the mysteries of the cosmos by probing it with his thought rather than relying on his senses. "The truth of a theory is in your mind," he once said, "not in your eyes."

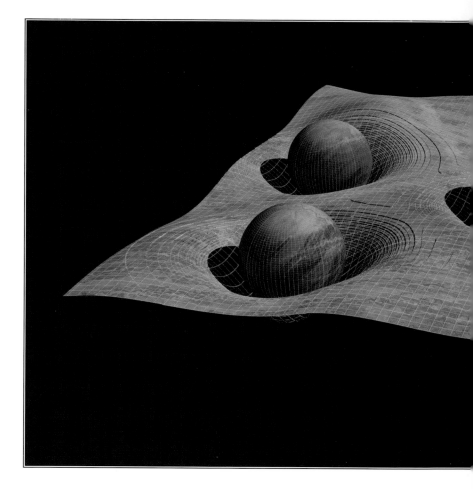

THE PRINCIPLE OF RELATIVITY

Translated by W. Perrett and G. B. Jeffery

ON THE ELECTRODYNAMICS OF MOVING BODIES

It is known that Maxwell's electrodynamics—as usually understood at the present time—when applied to moving bodies, leads to asymmetries which do not appear to be inherent in the phenomena. Take, for example, the reciprocal electrodynamic action of a magnet and a conductor. The observable phenomenon here depends only on the relative motion of the conductor and the magnet, whereas the customary view draws a sharp distinction between the two cases in which either the one or the other of these bodies is in motion. For if the magnet is in motion and

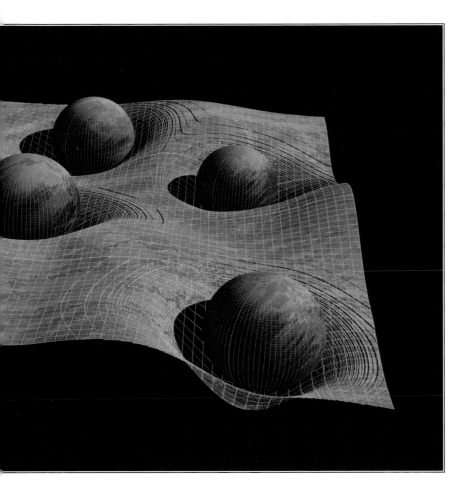

Einstein's view of the action of massive bodies warping the space-time continuum.

the conductor at rest, there arises in the neighborhood of the magnet an electric field with a certain definite energy, producing a current at the places where parts of the conductor are situated. But if the magnet is stationary and the conductor in motion, no electric field arises in the neighborhood of the magnet. In the conductor, however, we find an electromotive force, to which in itself there is no corresponding energy, but which gives rise—assuming equality of relative motion in the two cases discussed—to electric currents of the same path and intensity as those produced by the electric form in the former case.

Examples of this sort, together with the unsuccessful attempts to discover any motion of the Earth relatively to the "light medium," suggest that the phenomena of electrodynamics as well as of mechanics possess no properties corresponding to the idea of absolute rest. They suggest rather that, as has already been shown to the first order of small quantities, the same laws of electrodynamics and optics will be valid for all frames of reference for which the equations of mechanics hold good.[1] We will raise this conjecture (the purport of which will hereafter be called the "Principle of Relativity") to the status of a postulate, and also introduce another postulate, which is only apparently irreconcilable with the former, namely, that light is always propagated in empty space with a definite velocity c which is independent of the state of motion of the emitting body. These two postulates suffice for the attainment of a simple and consistent theory of the electrodynamics of moving bodies based on Maxwell's theory for stationary bodies. The introduction of a "luminiferous ether" will prove to be superfluous inasmuch as the view here to be developed will not require an "absolutely stationary space" provided with special properties, nor assign a velocity-vector to a point of the empty space in which electromagnetic processes take place.

The theory to be developed is based—like all electrodynamics—on the kinematics of the rigid body, since the assertions of any such theory have to do with the relationships between rigid bodies (systems of coordinates), clocks, and electromagnetic processes. Insufficient consideration of this circumstance lies at the root of the difficulties which the electrodynamics of moving bodies at present encounters.

I. KINEMATICAL PART

§ 1. DEFINITION OF SIMULTANEITY

Let us take a system of coordinates in which the equations of Newtonian mechanics hold good.[2] In order to render our presentation more precise and to distinguish this system of coordinates verbally from others which will be introduced hereafter, we call it the "stationary system."

If a material point is at rest relatively to this system of coordinates, its position can be defined relatively thereto by the employment of rigid standards of measurement and the methods of Euclidean geometry, and can be expressed in Cartesian coordinates.

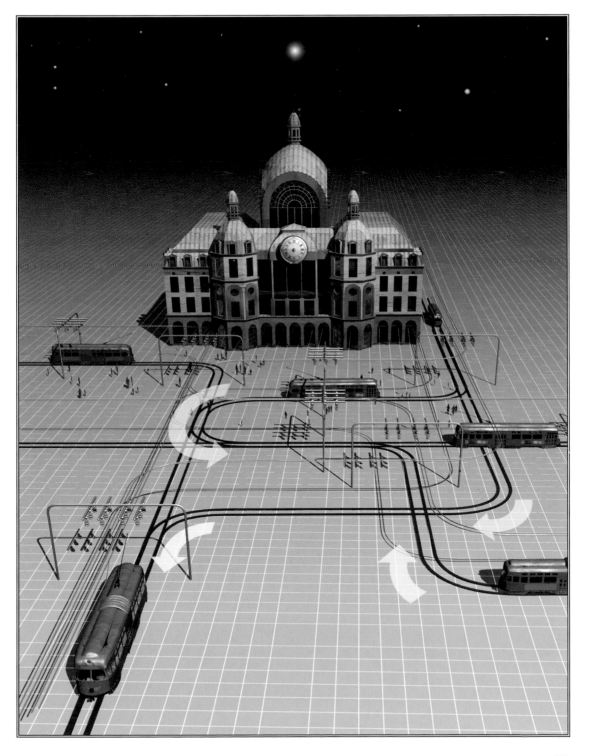

If we wish to describe the *motion* of a material point, we give the values of its coordinates as functions of the time. Now we must bear carefully in mind that a mathematical description of this kind has no physical meaning unless we are quite clear as to what we understand by "time." We have to take into account that all our judgments in which time plays a part are always judgments of *simultaneous events*. If, for instance, I say, "That train arrives here at 7 o'clock," I mean something like this: "The pointing of the small hand of my watch to 7 and the arrival of the train are simultaneous events."[3]

It might appear possible to overcome all the difficulties attending the definition of "time" by substituting "the position of the small hand of my watch" for "time." And in fact such a definition is satisfactory when we are concerned with defining a time exclusively for the place where the watch is located; but it is no longer satisfactory when we have to connect in time series of events occurring at different places, or—what comes to the same thing—to evaluate the times of events occurring at places remote from the watch.

We might, of course, content ourselves with time values determined by an observer stationed together with the watch at the origin of the coordinates, and coordinating the corresponding positions of the hands with light signals, given out by every event to be timed, and reaching him through empty space. But this coordination has the disadvantage that it is not independent of the standpoint of the observer with the watch or clock, as we know from experience. We arrive at a much more practical determination along the following line of thought.

If at the point A of space there is a clock, an observer at A can determine the time values of events in the immediate proximity of A by finding the positions of the hands, which are simultaneous with these events. If there is at the point B of space another clock in all respects resembling the one at A, it is possible for an observer at B to determine the time values of events in the immediate neighborhood of B. But it is not possible without further assumption to compare, in respect of time, an event at A with an event at B. We have so far defined only an "A time" and a "B time." We have not defined a common "time" for A and B, for the latter cannot be defined at all unless we establish *by definition* that the "time" required by light to travel from A to B equals the "time" it requires to travel from B to A. Let a ray of light start at the "A time" t_A from A towards B, let it at the "B time" t_B be reflected at B in the direction of A, and arrive again at A at the "A time" t'_A.

In accordance with definition the two clocks synchronize if

$$t_B - t_A = t'_A - t_B.$$

We assume that this definition of synchronism is free from contradictions, and possible for any number of points; and that the following relations are universally valid:

1. If the clock at B synchronizes with the clock at A, the clock at A synchronizes with the clock at B.

2. If the clock at A synchronizes with the clock at B and also with the clock at C, the clocks at B and C also synchronize with each other.

Thus with the help of certain imaginary physical experiments we have settled what is to be understood by synchronous stationary clocks located at different places, and have evidently obtained a definition of "simultaneous" or "synchronous," and of "time." The "time" of an event is that which is given simultaneously with the event by a stationary clock located at the place of the event, this clock being synchronous, and indeed synchronous for all time determinations, with a specified stationary clock.

In agreement with experience we further assume the quantity

$$\frac{2AB}{t'_A - t_A} = c$$

to be a universal constant—the velocity of light in empty space.

It is essential to have time defined by means of stationary clocks in the stationary system, and the time now defined being appropriate to the stationary system we call it "the time of the stationary system."

THE ILLUSTRATED ON THE SHOULDERS OF GIANTS

§ 2. ON THE RELATIVITY OF LENGTHS AND TIMES

The following reflections are based on the principle of relativity and on the principle of the constancy of the velocity of light. These two principles we define as follows:

1. The laws by which the states of physical systems undergo change are not affected, whether these changes of state be referred to the one or the other of two systems of coordinates in uniform translatory motion.

2. Any ray of light moves in the "stationary" system of coordinates with the determined velocity c, whether the ray be emitted by a stationary or by a moving body.

Hence

$$velocity = \frac{light\ path}{time\ interval}$$

where time interval is to be taken in the sense of the definition in § 1.

Let there be given a stationary rigid rod; and let its length be l as measured by a measuring-rod which is also stationary. We now imagine the axis of the rod lying along the axis of x of the stationary system of coordinates, and that a uniform motion of parallel translation with velocity v along the axis of x in the direction of increasing x is then imparted to the rod. We now inquire as to the length of the moving rod, and imagine its length to be ascertained by the following two operations:

(a) The observer moves together with the given measuring-rod and the rod to be measured, and measures the length of the rod directly by superposing the measuring-rod, in just the same way as if all three were at rest.

(b) By means of stationary clocks set up in the stationary system and synchronizing in accordance with § 1, the observer ascertains at what points of the stationary system the two ends of the rod to be measured are located at a definite time. The distance between these two points, measured by the measuring-rod already employed, which in this case is at rest, is also a length which may be designated "the length of the rod."

In accordance with the principle of relativity the length to be discovered by the operation (a)—we will call it "the length of the rod in the moving system"—must be equal to the length l of the stationary rod.

The length to be discovered by the operation (*b*) we will call "the length of the (moving) rod in the stationary system." This we shall determine on the basis of our two principles, and we shall find that it differs from *l*.

Current kinematics tacitly assumes that the lengths determined by these two operations are precisely equal, or in other words, that a moving rigid body at the epoch *t* may in geometrical respects be perfectly represented by *the same* body *at rest* in a definite position.

We imagine further that at the two ends A and B of the rod, clocks are placed which synchronize with the clocks of the stationary system, that is to say that their indications correspond at any instant to the "time of the stationary system" at the places where they happen to be. These clocks are therefore "synchronous in the stationary system."

We imagine further that with each clock there is a moving observer, and that these observers apply to both clocks the criterion established in § 1 for the synchronization of two clocks. Let a ray of light depart from A at the time[4] t_A, let it be reflected at B at the time t_B, and reach A again at the time t'_A. Taking into consideration the principle of the constancy of the velocity of light we find that

$$t_B - t_A = \frac{r_{AB}}{c - v} \quad \text{and} \quad t'_A - t_B = \frac{r_{AB}}{c + v}$$

where r_{AB} denotes the length of the moving rod—measured in the stationary system. Observers moving with the moving rod would thus find that the two clocks were not synchronous, while observers in the stationary system would declare the clocks to be synchronous.

So we see that we cannot attach any *absolute* signification to the concept of simultaneity, but that two events which, viewed from a system of coordinates, are simultaneous, can no longer be looked upon as simultaneous events when envisaged from a system which is in motion relatively to that system.

———————

ON THE INFLUENCE OF GRAVITATION ON THE PROPOGATION OF LIGHT

Translated from "Über den Einfluss der Schwerkraft auf die
Ausbreitung des Lichtes," Annalen der Physik, 35, 1911.

In a memoir published four years ago[5] I tried to answer the question
whether the propagation of light is influenced by gravitation. I return to
this theme, because my previous presentation of the subject does not satisfy me, and for a stronger reason, because I now see that one of the most
important consequences of my former treatment is capable of being tested
experimentally. For it follows from the theory here to be brought forward,
that rays of light, passing close to the Sun, are deflected by its gravitational
field, so that the angular distance between the Sun and a fixed star appearing near to it is apparently increased by nearly a second of arc.

In the course of these reflections further results are yielded which relate to gravitation. But as the exposition of the entire group of considerations would be rather difficult to follow, only a few quite elementary reflections will be given in the following pages, from which the reader will readily be able to inform himself as to the suppositions of the theory and its line of thought. The relations here deduced, even if the theoretical foundation is sound, are valid only to a first approximation.

§ I. A HYPOTHESIS AS TO THE PHYSICAL NATURE OF THE GRAVITATIONAL FIELD

In a homogeneous gravitational field (acceleration of gravity γ) let there be a stationary system of coordinates K, orientated so that the lines of force of the gravitational field run in the negative direction of the axis of z. In a space free of gravitational fields let there be a second system of coordinates K', moving with uniform acceleration (γ) in the positive direction of its axis of z. To avoid unnecessary complications, let us for the present disregard the theory of relativity, and regard both systems from the customary point of view of kinematics, and the movements occurring in them from that of ordinary mechanics.

Relatively to K, as well as relatively to K', material points which are not subjected to the action of other material points, move in keeping with the equations

$$\frac{d^2 x}{dt^2} = 0, \frac{d^2 y}{dt^2} = 0, \frac{d^2 z}{dt^2} = -\gamma.$$

For the accelerated system K' this follows directly from Galileo's principle, but for the system K, at rest in a homogeneous gravitational field, from the experience that all bodies in such a field are equally and uniformly accelerated. This experience, of the equal falling of all bodies in the gravitational field, is one of the most universal which the observation of nature has yielded; but in spite of that the law has not found any place in the foundations of our edifice of the physical universe.

But we arrive at a very satisfactory interpretation of this law of experience, if we assume that the systems K and K' are physically exactly equivalent, that is, if we assume that we may just as well regard the

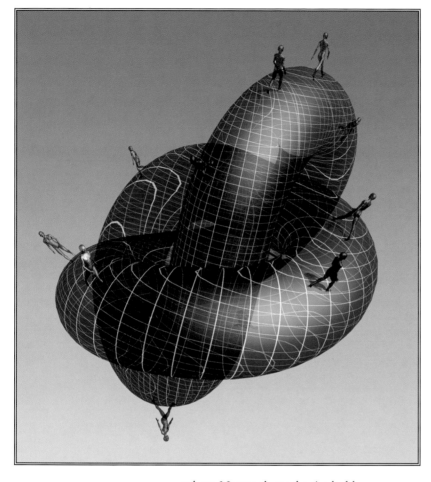

system K as being in a space free from gravitational fields, if we then regard K as uniformly accelerated. This assumption of exact physical equivalence makes it impossible for us to speak of the absolute acceleration of the system of reference, just as the usual theory of relativity forbids us to talk of the absolute velocity of a system;[6] and it makes the equal falling of all bodies in a gravitational field seem a matter of course.

As long as we restrict ourselves to purely mechanical processes in the realm where Newton's mechanics holds sway, we are certain of the equivalence of the systems K and K'. But this view of ours will not have any deeper significance unless the systems K and K' are equivalent with respect to all physical processes, that is, unless the laws of nature with respect to K are in entire agreement with those with respect to K'. By assuming this to be so, we arrive at a principle which, if it is really true, has great heuristic importance. For by theoretical consideration of processes which take place relatively to a system of reference with uniform acceleration, we obtain information as to the career of processes in a homogeneous gravitational field. We shall now show, first of all, from the standpoint of the ordinary theory of relativity, what degree of probability is inherent in our hypothesis.

§ 2. ON THE GRAVITATION OF ENERGY

One result yielded by the theory of relativity is that the inertia mass of a body increases with the energy it contains; if the increase of energy amounts to E, the increase in inertia mass is equal to E/c^2, when c denotes the velocity of light. Now is there an increase of gravitating mass corresponding to this increase of inertia mass? If not, then a body would fall in the same gravitational field with varying acceleration according to the energy it contained. That highly satisfactory result of the theory of relativity by which the law of the conservation of mass is merged in the law of conservation of energy could not be maintained, because it would compel us to abandon the law of the conservation of mass in its old form for inertia mass, and maintain it for gravitating mass.

But this must be regarded as very improbable. On the other hand, the usual theory of relativity does not provide us with any argument from which to infer that the weight of a body depends on the energy contained in it. But we shall show that our hypothesis of the equivalence of the systems K and K' gives us gravitation of energy as a necessary consequence.

Let the two material systems S_1 and S_2, provided with instruments of measurement, be situated on the z-axis of K at the distance h from each other[7], so that the gravitation potential in S_2 is greater than that in S_1 by γh. Let a definite quantity of energy E be emitted from S_2 towards S_1. Let the quantities of energy in S_1 and S_2 be measured by contrivances which—brought to one place in the system z and there compared—shall be perfectly alike. As to the process of this conveyance of energy by radiation we can make no *a priori* assertion, because we do not know the influence of the gravitational field on the radiation and the measuring instruments in S_1 and S_2.

But by our postulate of the equivalence of K and K' we are able, in place of the system K in a homogeneous gravitational field, to set the gravitation-free system K', which moves with uniform acceleration in the direction of positive z, and with the z-axis of which the material systems S_1 and S_2 are rigidly connected.

We judge of the process of the transference of energy by radiation

OPPOSITE PAGE

Einstein's theoretical model reveals that time and space are inseparable. While Newton's time was separate from space, as if it were a railroad track that stretched to infinity in both directions, Einstein's understanding was that his theory of relativity shows that time and space were inextricably interconnected.

One cannot curve space without involving time also. Thus time has a shape. Nevertheless, as shown opposite, time appears to have a one-way direction.

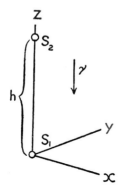

from S_2 to S_1 from a system K_0, which is to be free from acceleration. At the moment when the radiation energy E_2 is emitted from S_2 toward S_1, let the velocity of K' relatively to K_0 be zero. The radiation will arrive at S_1 when the time h/c has elapsed (to a first approximation). But at this moment the velocity of S_1 relatively to K_0 is $\gamma h/c = v$. Therefore by the ordinary theory of relativity the radiation arriving at S_1 does not possess the energy E_2, but a greater energy E_1, which is related to E_2, to a first approximation by the equation[8]

$$E_1 = E_2\left(1 + \frac{v}{c}\right) = E_2\left(1 + \gamma\frac{h}{c^2}\right) \tag{1}$$

By our assumption exactly the same relation holds if the same process takes place in the system K, which is not accelerated, but is provided with a gravitational field. In this case we may replace γh by the potential Φ of the gravitation vector in S_2, if the arbitrary constant of Φ in S_1 is equated to zero. We then have the equation

$$E_1 = E_2 + \frac{E_2}{c^2}\Phi \tag{1a}$$

This equation expresses the law of energy for the process under observation. The energy E_1 arriving at S_1 is greater than the energy E_2, measured by the same means, which was emitted in S_2, the excess being the potential energy of the mass E_2/c^2 in the gravitational field. It thus proves that for the fulfillment of the principle of energy we have to ascribe to the energy E, before its emission in S_2, a potential energy due to gravity, which corresponds to the gravitational mass E/c^2. Our assumption of the equivalence of K and K' thus removes the difficulty mentioned at the beginning of this paragraph which is left unsolved by the ordinary theory of relativity.

The meaning of this result is shown particularly clearly if we consider the following cycle of operations:

1. The energy E, as measured in S_2, is emitted in the form of radiation in S_2 towards S_1, where, by the result just obtained, the energy $E(1 + \gamma h/c^2)$, as measured in S_1, is absorbed.

2. A body W of mass M is lowered from S_2 to S_1, work $M\gamma h$ being done in theprocess.

3. The energy E is transferred from S_1 to the body W while W is in S_1. Let the gravitational mass M be thereby changed so that it acquires the value M'.

4. Let W be again raised to S_2, work $M'\gamma h$ being done in the process.

5. Let E be transferred from W back to S_2.

The effect of this cycle is simply that S_1 has undergone the increase of energy $E\gamma h/c^2$, and that the quantity of energy $M'\gamma k - M\gamma h$ has been conveyed to the system in the form of mechanical work. By the principle of energy, we must therefore have

$$E\gamma \frac{h}{c^2} = M'\gamma h - M\gamma h,$$

or

$$M' - M = E/c^2 \dots (1b)$$

The increase in gravitational mass is thus equal to E/c^2, and therefore equal to the increase in inertia mass as given by the theory of relativity.

The result emerges still more directly from the equivalence of the systems K and K', according to which the gravitational mass in respect of K is exactly equal to the inertia mass in respect of K'; energy must therefore possess a gravitational mass which is equal to its inertia mass. If a mass M_0 be suspended on a spring balance in the system K', the balance will indicate the apparent weight $M_0\gamma$ on account of the inertia of M_0. If the quantity of energy E be transferred to M_0, the spring balance, by the law of the inertia of energy, will indicate $(M_0 + E/c^2)\gamma$. By reason of our fundamental assumption exactly the same thing must occur when the experiment is repeated in the system K, that is, in the gravitational field.

3. TIME AND THE VELOCITY OF LIGHT IN THE GRAVITATIONAL FIELD

If the radiation emitted in the uniformly accelerated system K' in S_2 toward S_1 had the frequency ν_2 relatively to the clock in S_2, then, relatively to S_1, at its arrival in S_1 it no longer has the frequency ν_2, relatively to an identical clock in S_1, but a greater frequency ν_1, such that to a first approximation

$$\nu_1 = \nu_2 \left(1 + \gamma \frac{h}{c^2} \right). \tag{2}$$

For if we again introduce the unaccelerated system of reference K_0, relatively to which, at the time of the emission of light, K' has no velocity, then S_1, at the time of arrival of the radiation at S_1, has, relatively to K_0, the velocity $\gamma h/c$, from which, by Doppler's principle, the relation as given results immediately.

In agreement with our assumption of the equivalence of the systems K' and K, this equation also holds for the stationary system of coordinates K, provided with a uniform gravitational field, if in it the transference by radiation takes place as described. It follows, then, that a ray of light emitted in S_2 with a definite gravitational potential, and possessing at its emission the frequency ν_2—compared with a clock in S_2—will, at its arrival in S_1, possess a different frequency ν_1—measured by an identical clock in S_1. For γh we substitute the gravitational potential Φ of S_2—that of S_1 being taken as zero—and assume that the relation which we have deduced for the homogeneous gravitational field also holds for other forms of field. Then

$$\nu_1 = \nu_2 \left(1 + \frac{\Phi}{c^2} \right) \tag{2a}$$

This result (which by our deduction is valid to a first approximation) permits, in the first place, of the following application. Let ν_0 be the vibration-number of an elementary light-generator, measured by a delicate clock at the same place. Let us imagine them both at a place on the surface of the Sun (where our S_2 is located). Of the light there emitted, a portion reaches the Earth (S_1), where we measure the frequency of the

arriving light with a clock U in all respects resembling the one just mentioned. Thenby (2a),

$$v = v_0\left(1 + \frac{\Phi}{c^2}\right),$$

*Is time reversible? It would
appear there are few arguments
for it and a cosmos against it.*

where Φ is the (negative) difference of gravitational potential between
the surface of the Sun and the Earth. Thus according to our view the
spectral lines of sunlight, as compared with the corresponding spectral
lines of terrestrial sources of light, must be somewhat displaced toward
the red, in fact by the relative amount

$$\frac{\nu_0 - \nu}{\nu_0} = -\frac{\Phi}{c^2} = 2.10^{-6}$$

If the conditions under which the solar bands arise were exactly known, this shifting would be susceptible of measurement. But as other influences (pressure, temperature) affect the position of the centers of the spectral lines, it is difficult to discover whether the inferred influence of the gravitational potential really exists.[9]

On a superficial consideration equation (2), or (2a), respectively, seems to assert an absurdity. If there is constant transmission of light from S_2 to S_1, how can any other number of periods per second arrive in S_1 than is emitted in S_2? But the answer is simple. We cannot regard ν_2 or respectively ν_1 simply as frequencies (as the number of periods per second) since we have not yet determined the time in system K. What ν_2 denotes is the number of periods with reference to the time-unit of the clock U in S_2, while n1 denotes the number of periods per second with reference to the identical clock in S_1. Nothing compels us to assume that the clocks U in different gravitation potentials must be regarded as going at the same rate. On the contrary, we must certainly define the time in K in such a way that the number of wave crests and troughs between S_2 and S_1 is independent of the absolute value of time; for the process under observation is by nature a stationary one. If we did not satisfy this condition, we should arrive at a definition of time by the application of which time would merge explicitly into the laws of nature, and this would certainly be unnatural and unpractical. Therefore the two clocks in S_1 and S_2 do not both give the "time" correctly. If we measure time in S_1 with the clock U, then we must measure time in S_2 with a clock which goes $1 + M/c^2$ times more slowly than the clock U when compared with U at one and the same place. For when measured by such a clock the frequency of the ray of light which is considered above is at its emission in S_2

$$\nu_2\left(1 + \frac{\Phi}{c^2}\right)$$

and is therefore, by (2a), equal to the frequency ν_1 of the same ray of light on its arrival in S_1.

This has a consequence which is of fundamental importance for our theory. For if we measure the velocity of light at different places in the accelerated, gravitation free system K', employing clocks U of identical constitution, we obtain the same magnitude at all these places. The same holds good, by our fundamental assumption, for the system K as well. But from what has just been said we must use clocks of unlike constitution, for measuring time at places with differing gravitation potential. For measuring time at a place which, relatively to the origin of the coordinates, has the gravitation potential Φ, we must employ a clock which—when removed to the origin of coordinates —goes $(1 + \Phi/c^2)$ times more slowly than the clock used for measuring time at the origin of coordinates. If we call the velocity of light at the origin of coordinates c_0, then the velocity of light c at a place with the gravitation potential Φ will be given by the relation

$$c = c_0\left(1 + \frac{\Phi}{c^2}\right) \tag{3}$$

The principle of the constancy of the velocity of light holds good according to this theory in a different form from that which usually underlies the ordinary theory of relativity.

4. BENDING OF LIGHT-RAYS IN THE GRAVITATIONAL FIELD

From the proposition which has just been proved, that the velocity of light in the gravitational field is a function of the place, we may easily infer, by means of Huyghens's principle, that light-rays propagated across a gravitational field undergo deflection. For let E be a wave front of a plane light-wave at the time t, and let P_1 and P_2 be two points in that plane at

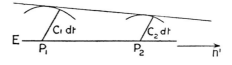

unit distance from each other. P_1 and P_2 lie in the plane of the paper, which is chosen so that the differential coefficient of Φ, taken in the direction of the normal to the plane, vanishes, and therefore also that of c. We obtain the corresponding wave front at time $t + dt$, or, rather, its line of section with the plane of the paper, by describing circles round the points P_1 and P_2 with radii $c_1 dt$ and $c_2 dt$ respectively, where c_1 and c_2 denote the velocity of light at the points P_1 and P_2 respectively, and by drawing the tangent to these circles. The angle through which the light-ray is deflected in the path cdt is therefore

$$\left(c_1 - c_2\right)dt = -\frac{\partial c}{\partial n'}dt,$$

if we calculate the angle positively when the ray is bent toward the side of increasing n'. The angle of deflection per unit of path of the light-ray is thus

$$-\frac{1}{c}\frac{\partial c}{\partial n'} \quad \text{or by (3)} \quad -\frac{1}{c^2}\frac{\partial \Phi}{\partial n'}$$

Finally, we obtain for the deflection which a light-ray experiences toward the side n' on any path (s) the expression

$$a = -\frac{1}{c^2}\int\frac{\partial \Phi}{\partial n'}ds \qquad (4)$$

The standard model of the life and death of our universe. Without Einstein's theoretical work this model would not have been mathematically possible.

From left to right in this illustration—trillionths of a second after the Big Bang, the universe inflates from smaller than an atom with the mass of a bag of sugar to the size of a galaxy.
The universe continues to expand with galaxies and eventually, stars, atoms, and particles becoming further apart until the whole universe is an exhausted and barren void. A second model suggests that the acceleration finally ceases and the universe collapses under gravitational forces into a vast black hole and the Big Crunch.

We might have obtained the same result by directly considering the propagation of a ray of light in the uniformly accelerated system K', and transferring the result to the system K, and thence to the case of a gravitational field of any form.

By equation (4) a ray of light passing along by a heavenly body suffers a deflection to the side of the diminishing gravitational potential, that is, on the side directed toward the heavenly body, of the magnitude

$$a = \frac{1}{c^2} \int_{\theta = -\frac{1}{2}\pi}^{\theta = \frac{1}{2}\pi} \frac{k\mathrm{M}}{r^2} \cos \theta ds = 2\frac{k\mathrm{M}}{c^2 \Delta}$$

where k denotes the constant of gravitation, M the mass of the heavenly body, Δ the distance of the ray from the center of the body. A ray of light going past the Sun would accordingly undergo deflection to the amount of $4 \cdot 10^{-6}$ = .83 seconds of arc. The angular distance of the star from the center of the Sun appears to be increased by this amount. As the fixed stars in the parts of the sky near the Sun are visible during total eclipses of the Sun, this consequence of the theory may be compared with experience. With the planet Jupiter the displacement to be expected reaches to about 1/100 of the amount given. It would be a most desirable thing if astronomers would take up the question here raised. For apart from any theory there is the question whether it is possible with the equipment at present available to detect an influence of gravitational fields on the propagation of light.

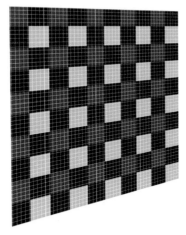
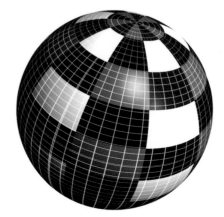

THE FOUNDATION OF
THE GENERAL THEORY OF RELATIVITY

Translated from "Die Grundlage der allgemeinen
Relativitätstheorie," Annalen der Physik, 49, 1916.

A. FUNDAMENTAL CONSIDERATIONS ON THE POSTULATE OF RELATIVITY

§ 1. OBSERVATIONS ON THE SPECIAL THEORY OF RELATIVITY

The special theory of relativity is based on the following postulate, which
is also satisfied by the mechanics of Galileo and Newton.

If a system of coordinates K is chosen so that, in relation to it, physical laws hold good in their simplest form, the *same* laws also hold good
in relation to any other system of coordinates K' moving in uniform
translation relatively to K. This postulate we call the "special principle of
relativity." The word "special" is meant to intimate that the principle is
restricted to the case when K' has a motion of uniform translation relatively to K, but that the equivalence of K' and K does not extend to the
case of non-uniform motion of K' relatively to K.

Thus the special theory of relativity does not depart from classical
mechanics through the postulate of relativity, but through the postulate

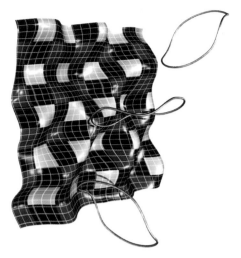

of the constancy of the velocity of light in vacuo, from which, in combination with the special principle of relativity, there follow, in the well-known way, the relativity of simultaneity, the Lorentzian transformation, and the related laws for the behavior of moving bodies and clocks.

The modification to which the special theory of relativity has subjected the theory of space and time is indeed far-reaching, but one important point has remained unaffected. For the laws of geometry, even according to the special theory of relativity, are to be interpreted directly as laws relating to the possible relative positions of solid bodies at rest; and, in a more general way, the laws of kinematics are to be interpreted as laws which describe the relations of measuring bodies and clocks. To two selected material points of a stationary rigid body there always corresponds a distance of quite definite length, which is independent of the locality and orientation of the body, and is also independent of the time. To two selected positions of the hands of a clock at rest relatively to the privileged system of reference there always corresponds an interval of time of a definite length, which is independent of place and time. We shall soon see that the general theory of relativity cannot adhere to this simple physical interpretation of space and time.

ABOVE (BOTH PAGES)

The theoretical histories of the universe—the flat membrane at far left indicates a need to specify boundaries, such as was the view of Earth when it was believed to be flat. If the universe goes off to infinity like a saddle (above left) one has the problem, again, of specifying the boundary conditions at infinity. If all the histories of the universe in imaginary time are closed surfaces such as those of the Earth, there is no need to specify boundary conditions at all. Beyond Einstein, with modern string theory (above right) we conceive of multiple dimensions within a membrane world.

227

§ 2. THE NEED FOR AN EXTENSION OF THE POSTULATE OF RELATIVITY

In classical mechanics, and no less in the special theory of relativity, there is an inherent epistemological defect which was, perhaps for the first time, clearly pointed out by Ernst Mach. We will elucidate it by the following example: Two fluid bodies of the same size and nature hover freely in space at so great a distance from each other and from all other masses that only those gravitational forces need be taken into account which arise from the interaction of different parts of the same body. Let the distance between the two bodies be invariable, and in neither of the bodies let there be any relative movements of the parts with respect to one another. But let either mass, as judged by an observer at rest relatively to the other mass, rotate with constant angular velocity about the line joining the masses. This is a verifiable relative motion of the two bodies. Now let us imagine that each of the bodies has been surveyed by means of measuring instruments at rest relatively to itself, and let the surface of S_1 prove to be a sphere, and that of S_2 an ellipsoid of revolution. Thereupon we put the question—What is the reason for this difference in the two bodies? No answer can be admitted as epistemologically satisfactory,[10] unless the reason given is an *observable fact of experience*. The law of causality has not the significance of a statement as to the world of experience, except when *observable facts* ultimately appear as causes and effects.

Newtonian mechanics does not give a satisfactory answer to this question. It pronounces as follows: The laws of mechanics apply to the space R_1, in respect to which the body S_1 is at rest, but not to the space R_2 in respect to which the body S_2 is at rest. But the privileged space R_1 of Galileo, thus introduced, is a merely *factitious* cause, and not a thing that can be observed. It is therefore clear that Newton's mechanics does not really satisfy the requirement of causality in the case under consideration, but only apparently does so, since it makes the factitious cause R_1 responsible for the observable difference in the bodies S_1 and S_2.

The only satisfactory answer must be that the physical system consisting of S_1 and S_2 reveals within itself no imaginable cause to which the differing behavior of S_1 and S_2 can be referred. The cause must therefore lie *outside* this system. We have to take it that the general laws of motion, which in particular determine the shapes of S_1 and S_2, must be such that the mechanical behavior of S_1 and S_2 is partly conditioned, in quite essential respects, by distant masses which we have not included in the system under consideration. These distant masses and their motions relative to S_1 and S_2 must then be regarded as the seat of the causes (which must be susceptible to observation) of the different behavior of our two bodies S_1 and S_2. They take over the rôle of the factitious cause R_1. Of all imaginable spaces R_1, R_2, etc., in any kind of motion relatively to one another, there is none which we may look upon as privileged a priori without reviving the above-mentioned epistemological objection. *The laws of physics must be of such a nature that they apply to systems of reference in any kind of motion.* Along this road we arrive at an extension of the postulate of relativity.

In addition to this weighty argument from the theory of knowledge, there is a well known physical fact which favors an extension of the theory of relativity. Let K be a Galilean system of reference, *i.e.* a system relatively to which (at least in the four-dimensional region under consideration) a mass, sufficiently distant from other masses, is moving with uniform motion in a straight line. Let K' be a second system of reference which is moving relatively to K in *uniformly accelerated* translation. Then, relatively to K', a mass sufficiently distant from other masses would have an accelerated motion such that its acceleration and direction of acceleration are independent of the material composition and physical state of the mass.

Relativity depends upon the constant of the speed of light (186,000 miles or 300,000 kilometers per second.) In a year it travels 5.6 trillion miles. This is a light year. It equals 63.240 astronomical units (the distance of Earth from the Sun). Pluto, our most distant planet in the solar system, is 49.3 astronomical units away, while the nearest star or sun, Alpha Centauri, is 4.3 light years from us. The edge of the Milky Way, our own galaxy, is fifty thousand light years away, while the nearest galaxy, Andromeda, is 2.3 million light years away. Most of the stars we can see with the naked eye are no more than a thousand light years away.

Does this permit an observer at rest relatively to K' to infer that he is on a "really" accelerated system of reference? The answer is in the negative; for the above-mentioned relation of freely movable masses to K' may be interpreted equally well in the following way. The system of reference K' is unaccelerated, but the space-time territory in question is under the sway of a gravitational field, which generates the accelerated motion of the bodies relatively to K'.

This view is made possible for us by the teaching of experience as to the existence of a field of force, namely, the gravitational field, which possesses the remarkable property of imparting the same acceleration to all bodies.[11] The mechanical behavior of bodies relatively to K' is the same as presents itself to experience in the case of systems which we are wont to regard as "stationary" or as "privileged." Therefore, from the physical standpoint, the assumption readily suggests itself that the systems K and K' may both with equal right be looked upon as "stationary," that is to say, they have an equal title as systems of reference for the physical description of phenomena.

It will be seen from these reflections that in pursuing the general theory of relativity we shall be led to a theory of gravitation, since we are able to "produce" a gravitational field merely by changing the system of coordinates. It will also be obvious that the principle of the constancy of the velocity of light *in vacuo* must be modified, since we easily recognize that the path of a ray of light with respect to K' must in general be curvilinear, if with respect to K light is propagated in a straight line with a definite constant velocity.

§ 3. THE SPACE-TIME CONTINUUM. REQUIREMENT OF GENERAL CO-VARIANCE FOR THE EQUATIONS EXPRESSING GENERAL LAWS OF NATURE

In classical mechanics, as well as in the special theory of relativity, the coordinates of space and time have a direct physical meaning. To say that a point-event has the X1 coordinate x1 means that the projection of the point-event on the axis of X1, determined by rigid rods and in accordance with the rules of Euclidean geometry, is obtained by measuring off a given rod (the unit of length) x1 times from the origin of coordinates along the axis of X1. To say that a point-event has the X4 coordinates $x4 = t$, means that a standard clock, made to measure time in a definite unit period, and which is stationary relatively to the system of coordinates and practically coincident in space with the point-event,[12] will have measured off $x4 = t$ periods at the occurrence of the event.

This view of space and time has always been in the minds of physicists, even if, as a rule, they have been unconscious of it. This is clear from the part which these concepts play in physical measurements; it must also have underlain the reader's reflections on the preceding paragraph (§ 2) for him to connect any meaning with what he there read. But we shall now show that we must put it aside and replace it by a more general view, in order to be able to carry through the postulate of general relativity, if the special theory of relativity applies to the special case of the absence of a gravitational field.

In a space which is free of gravitational fields we introduce a Galilean system of reference K (x, y, z, t), and also a system of coordinates K' (x', y', z', t') in uniform rotation relatively to K. Let the origins of both systems, as well as their axes of Z, permanently coincide. We shall show that for a space-time measurement in the system K' the above definition of the physical meaning of lengths and times cannot be maintained. For reasons

THE ILLUSTRATED ON THE SHOULDERS OF GIANTS

OPPOSITE PAGE

Three models of universes: their inflation, expansion, and contraction.

TOP

A universe that has a sudden expansion but falls back on itself to create a Big Crunch with a massive black hole.

MIDDLE

A universe that appears to be like our own in which there is a second accelerated expansion that could continue until the universe becomes a lifeless exhausted void or like the top end in a black hole.

BOTTOM

A universe that expands early in its life and continues without managing to create galactic systems or major stars. The orange circle in each illustration signifies the point at which the major accelerated expansion occurs.

of symmetry it is clear that a circle around the origin in the X, Y plane of K may at the same time be regarded as a circle in the X', Y' plane of K'. We suppose that the circumference and diameter of this circle have been measured with a unit measure infinitely small compared with the radius, and that we have the quotient of the two results. If this experiment were performed with a measuring-rod at rest relatively to the Galilean system K, the quotient would be B. With a measuring-rod at rest relatively to K', the quotient would be greater than B. This is readily understood if we envisage the whole process of measuring from the "stationary" system K, and take into consideration that the measuring-rod applied to the periphery undergoes a Lorentzian contraction, while the one applied along the radius does not. Hence Euclidean geometry does not apply to K'. The notion of coordinates defined above, which presupposes the validity of Euclidean geometry, therefore breaks down in relation to the system K'. So, too, we are unable to introduce a time corresponding to physical requirements in K', indicated by clocks at rest, relatively to K'. To convince ourselves of this impossibility, let us imagine two clocks of identical constitution placed, one at the origin of coordinates, and the other at the circumference of the circle, and both envisaged from the "stationary" system K. By a familiar result of the special theory of relativity, the clock at the circumference—judged from K—goes more slowly than the other, because the former is in motion and the latter at rest. An observer at the common origin of coordinates, capable of observing the clock at the circumference by means of light, would therefore see it lagging behind the clock beside him. As he will not make up his mind to let the velocity of light along the path in question depend explicitly on the time, he will interpret his observations as showing that the clock at the circumference "really" goes more slowly than the clock at the origin. So he will be obliged to define time in such a way that the rate of a clock depends upon where the clock may be.

We therefore reach this result: In the general theory of relativity, space and time cannot be defined in such a way that differences of the spatial coordinates can be directly measured by the unit measuring-rod, or differences in the time coordinate by a standard clock.

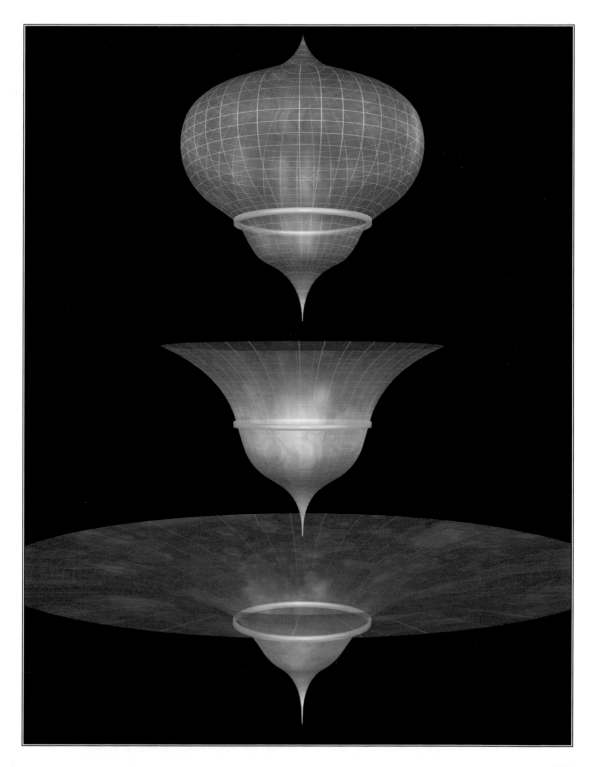

OPPOSITE PAGE

Wormholes connecting across space and time. The danger in theory is that they only remain open a short time before severing the bridge.

The method hitherto employed for laying coordinates into the space-time continuum in a definite manner thus breaks down, and there seems to be no other way which would allow us to adapt systems of coordinates to the four-dimensional universe so that we might expect from their application a particularly simple formulation of the laws of nature. So there is nothing for it but to regard all imaginable systems of coordinates, on principle, as equally suitable for the description of nature. This comes to requiring that:

The general laws of nature are to be expressed by equations which hold good for all systems of coordinates, that is, are covariant with respect to any substitutions whatever (generally covariant).

It is clear that a physical theory which satisfies this postulate will also be suitable for the general postulate of relativity. For the sum of *all* substitutions in any case includes those which correspond to all relative motions of three-dimensional systems of coordinates. That this requirement of general covariance, which takes away from space and time the last remnant of physical objectivity, is a natural one, will be seen from the following reflection. All our space-time verifications invariably amount to a determination of space-time coincidences. If, for example, events consisted merely in the motion of material points, then ultimately nothing would be observable but the meetings of two or more of these points. Moreover, the results of our measurings are nothing but verifications of such meetings of the material points of our measuring instruments with other material points, coincidences between the hands of a clock and points on the clock dial, and observed point-events happening at the same place at the same time.

The introduction of a system of reference serves no other purpose than to facilitate the description of the totality of such coincidences. We allot to the universe four space-time variables x_1, x_2, x_3, x_4 in such a way that for every point-event there is a corresponding system of values of the variables $x_1 \ldots x_4$. To two coincident point-events there corresponds one system of values of the variables $x_1 \ldots x_4$, i.e. coincidence is characterized by the identity of the coordinates. If, in place of the variables $x_1 \ldots x_4$, we introduce functions of them, x'_1, x'_2, x'_3, x'_4, as a new system of coordinates, so that the systems of values are made to correspond to one another without ambiguity, the equality of all four coordinates in the new system will also serve as an expression for the space-time coincidence of the two point-events. As all our physical experience can be ultimately reduced to such coincidences, there is no immediate reason for preferring certain systems of coordinates to others, that is to say, we arrive at the requirement of general covariance.

———

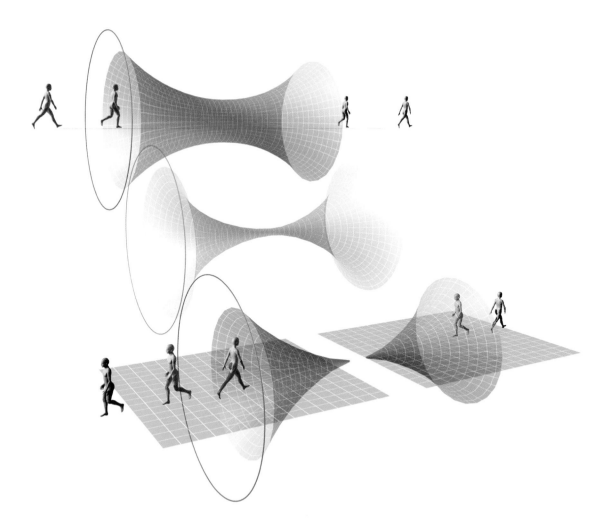

COSMOLOGICAL CONSIDERATIONS ON THE GENERAL THEORY OF RELATIVITY

Translated from "Kosmologische Betrachtungen zur allgemeinen Relativitätstheorie," Sitzungsberichte der Preussischen Akad. d. Wissenschaften, 1917.

It is well known that Poisson's equation

$$\nabla^2 \phi = 4\pi K \rho \tag{1}$$

in combination with the equations of motion of a material point is not as yet a perfect substitute for Newton's theory of action at a distance. There is still to be taken into account the condition that at spatial infinity

the potential ϕ tends toward a fixed limiting value. There is an analogous state of things in the theory of gravitation in general relativity. Here, too, we must supplement the differential equations by limiting conditions at spatial infinity, if we really have to regard the universe as being of infinite spatial extent.

In my treatment of the planetary problem I chose these limiting conditions in the form of the following assumption: it is possible to select a system of reference so that at spatial infinity all the gravitational potentials $g_{\upsilon\upsilon}$ become constant. But it is by no means evident *a priori* that we may lay down the same limiting conditions when we wish to take larger portions of the physical universe into consideration. In the following pages the reflections will be given which, up to the present, I have made on this fundamentally important question.

The paradox of wormholes brings up the notion that if we travel back in time we have the power to alter the past, and, therefore, the future too. What happens if you can go back in time and kill your grandfather before your father or mother was conceived?

§ 1. THE NEWTONIAN THEORY

It is well known that Newton's limiting condition of the constant limit for ϕ at spatial infinity leads to the view that the density of matter becomes zero at infinity. For we imagine that there may be a place in universal space round about which the gravitational field of matter, viewed on a large scale, possesses spherical symmetry. It then follows from Poisson's equation that, in order that ϕ may tend to a limit at infinity, the mean density ρ must decrease toward zero more rapidly than $1/r^2$ as the distance r from the center increases.[13] In this sense, therefore, the universe according to Newton is finite, although it may possess an infinitely great total mass.

From this it follows in the first place that the radiation emitted by the heavenly bodies will, in part, leave the Newtonian system of the universe, passing radially outwards, to become ineffective and lost in the infinite. May not entire heavenly bodies fare likewise? It is hardly possible to give a negative answer to this question. For it follows from the assumption of a finite limit for ϕ at spatial infinity that a heavenly body with finite kinetic energy is able to reach spatial infinity by overcoming the Newtonian forces of attraction. By statistical mechanics this case must occur from time to time, as long as the total energy of the stellar system—transferred to one single star—is great enough to send that star on its journey to infinity, whence it never can return.

We might try to avoid this peculiar difficulty by assuming a very high value for the limiting potential at infinity. That would be a possible way, if the value of the gravitational potential were not itself necessarily conditioned by the heavenly bodies. The truth is that we are compelled to regard the occurrence of any great differences of potential of the gravitational field as contradicting the facts. These differences must really be of so low an order of magnitude that the stellar velocities generated by them do not exceed the velocities actually observed.

If we apply Boltzmann's law of distribution for gas molecules to the stars, by comparing the stellar system with a gas in thermal equilibrium, we find that the Newtonian stellar system cannot exist at all. For there is a finite ratio of densities corresponding to the finite difference of potential between the center and spatial infinity. A vanishing of the density at infinity thus implies a vanishing of the density at the center.

It seems hardly possible to surmount these difficulties on the basis of the Newtonian theory. We may ask ourselves the question whether they can be removed by a modification of the Newtonian theory. First of all we will indicate a method which does not in itself claim to be taken seriously; it merely serves as a foil for what is to follow. In place of Poisson's equation we write

$$\nabla^2 \phi - \lambda \phi = 4\pi \kappa \rho \tag{2}$$

where λ denotes a universal constant. If ρ_0 be the uniform density of distribution of mass, then

$$\phi = -\frac{4\pi\kappa}{\lambda}\rho_0 \tag{3}$$

is a solution of equation (2). This solution would correspond to the case in which the matter of the fixed stars was distributed uniformly through space, if the density ρ_0 is equal to the actual mean density of the matter in the universe. The solution then corresponds to an infinite extension of the central space, filled uniformly with matter. If, without making any change in the mean density, we imagine matter to be non-uniformly distributed locally, there will be, over and above the ϕ with the constant value of equation (3), an additional ϕ, which in the neighborhood of denser masses will so much the more resemble the Newtonian field as $\lambda\phi$ is smaller in comparison with $4\pi\kappa\rho$.

A universe so constituted would have, with respect to its gravitational field, no center. A decrease of density in spatial infinity would not have to be assumed, but both the mean potential and mean density would remain constant to infinity. The conflict with statistical mechanics which we found in the case of the Newtonian theory is not repeated. With a definite but extremely small density, matter is in equilibrium, without any internal material form (pressures) being required to maintain equilibrium.

A Star in its stable stage showing light escaping from its surface

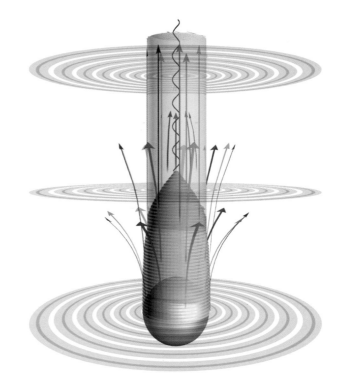

A Star begins to collapse (mid stage) and the light is pulled back to its surface until a point arrives (the event horizon) when no light will escape. The star becomes a singularity.

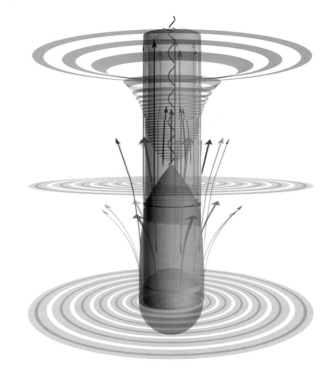

§ 2. THE BOUNDARY CONDITIONS ACCORDING TO THE GENERAL THEORY OF RELATIVITY

In the present paragraph I shall conduct the reader over the road that I have myself traveled, rather a rough and winding road, because otherwise I cannot hope that he will take much interest in the result at the end of the journey. The conclusion I shall arrive at is that the field equations of gravitation which I have championed hitherto still need a slight modification, so that on the basis of the general theory of relativity those fundamental difficulties may be avoided which have been set forth in § 1 as confronting the Newtonian theory. This modification corresponds perfectly to the transition from Poisson's equation (1) to equation (2) of § 1. We finally infer that boundary conditions in spatial infinity fall away altogether, because the universal continuum in respect of its spatial dimensions is to be viewed as a self-contained continuum of finite spatial (three dimensional) volume.

The opinion which I entertained until recently, as to the limiting conditions to be laid down in spatial infinity, took its stand on the following considerations. In a consistent theory of relativity there can be no inertia *relatively* to "*space,*" but only an inertia of masses *relatively to one another.* If, therefore, I have a mass at a sufficient distance from all other masses in the universe, its inertia must fall to zero. We will try to formulate this condition mathematically.

According to the general theory of relativity the negative momentum is given by the first three components, the energy by the last component of the covariant tensor multiplied by $\sqrt{-g}$

$$m\sqrt{-g}\, g_{\mu\alpha}\, \frac{dx_\alpha}{ds} \tag{4}$$

where, as always, we set

$$ds^2 = g_{\mu\nu}dx_\mu dx_\nu \tag{5}$$

In the particularly perspicuous case of the possibility of choosing the system of coordinates so that the gravitational field at every point is spatially isotropic, we have more simply

$$ds^2 = -A\left(dx_1^2 + dx_2^2 + dx_3^2\right) + B dx_4^2$$

If, moreover, at the same time

$$\sqrt{-g} = 1 = \sqrt{A^3 B}$$

we obtain from (4), to a first approximation for small velocities,

$$m \frac{A}{\sqrt{B}} \frac{dx_1}{dx_4}, m \frac{A}{\sqrt{B}} \frac{dx_2}{dx_4}, m \frac{A}{\sqrt{B}} \frac{dx_3}{dx_4}$$

for the components of momentum, and for the energy (in the static case)

$$m\sqrt{B}.$$

From the expressions for the momentum, it follows that $m \frac{A}{\sqrt{B}}$ plays the part of the rest mass. As m is a constant peculiar to the point of mass, independently of its position, this expression, if we retain the condition $\sqrt{g} = 1$ at spatial infinity, can vanish only when A diminishes to zero, while B increases to infinity. It seems, therefore, that such a degeneration of the coefficients $g_{\mu\nu}$ is required by the postulate of relativity of all inertia. This requirement implies that the potential energy $m\sqrt{B}$ becomes infinitely great at infinity. Thus a point of mass can never leave the system; and a more detailed investigation shows that the same thing applies to light-rays. A system of the universe with such behavior of the gravitational potentials at infinity would not therefore run the risk of wasting away which was mooted just now in connection with the Newtonian theory.

I wish to point out that the simplifying assumptions as to the gravitational potentials on which this reasoning is based, have been introduced merely for the sake of lucidity. It is possible to find general formulations for the behavior of the $g_{\mu\nu}$ at infinity which express the essentials of the question without further restrictive assumptions.

At this stage, with the kind assistance of the mathematician J. Grommer, I investigated centrally symmetrical, static gravitational fields, degenerating at infinity in the way mentioned. The gravitational potentials $g_{\mu\nu}$ were applied, and from them the energy-tensor $T_{\mu\nu}$ of matter was calculated on the basis of the field equations of gravitation. But here it proved that for the system of the fixed stars no boundary conditions of the kind can come into question at all, as was also rightly emphasized by the astronomer de Sitter recently.

For the contravariant energy-tensor $T_{\mu\nu}$ of ponderable matter is given by

$$T^{\mu\nu} = \rho \frac{dx_\mu}{ds} \frac{dx_\nu}{ds},$$

where ρ is the density of matter in natural measure. With an appropriate choice of the system of coordinates the stellar velocities are very small in comparison with that of light. We may, therefore, substitute $\sqrt{g_{44}}\, dx_4$ for ds. This shows us that all components of $T^{\mu\nu}$ must be very small in comparison with the last component T^{44}. But it was quite impossible to reconcile this condition with the chosen boundary conditions. In the retrospect this result does not appear astonishing. The fact of the small velocities of the stars allows the conclusion that wherever there are fixed stars, the gravitational potential (in our case $\sqrt{B}$) can never be much greater than here on Earth. This follows from statistical reasoning, exactly as in the case of the Newtonian theory. At any rate, our calculations have convinced me that such conditions of degeneration for the $g_{\mu\nu}$ in spatial infinity may not be postulated.

After the failure of this attempt, two possibilities present themselves.

(a) We may require, as in the problem of the planets, that, with a suitable choice of the system of reference, the gmn in spatial infinity approximate to the values

$$\begin{array}{cccc} -1 & 0 & 0 & 0 \\ 0 & -1 & 0 & 0 \\ 0 & 0 & -1 & 0 \\ 0 & 0 & 0 & 1 \end{array}$$

(b) We may refrain entirely from laying down boundary conditions for spatial infinity claiming general validity; but at the spatial limit of the domain under consideration we have to give the $g_{\mu\nu}$ separately in each individual case, as hitherto we were accustomed to give the initial conditions for time separately.

The possibility (*b*) holds out no hope of solving the problem, but amounts to giving it up. This is an incontestable position, which is taken up at the present time by de Sitter.[14] But I must confess that such a complete resignation in this fundamental question is for me a difficult thing. I should not make up my mind to it until every effort to make headway toward a satisfactory view had proved to be vain.

Possibility (*a*) is unsatisfactory in more respects than one. In the first place those boundary conditions presuppose a definite choice of the system of reference, which is contrary to the spirit of the relativity principle. Secondly, if we adopt this view, we fail to comply with the requirement of the relativity of inertia. For the inertia of a material point of mass m (in natural measure) depends upon the $g_{\mu\nu}$; but these differ but little from their postulated values, as given above, for spatial infinity. Thus inertia would indeed be *influenced*, but would not be *conditioned* by matter (present in finite space). If only one single point of mass were present, according to this view, it would possess inertia, and in fact an inertia almost as great as when it is surrounded by the other masses of the actual universe. Finally, those statistical objections must be raised against this view which were mentioned in respect of the Newtonian theory.

From what has now been said it will be seen that I have not succeeded in formulating boundary conditions for spatial infinity. Nevertheless, there is still a possible way out, without resigning as suggested under (*b*). For if it were possible to regard the universe as a continuum which is *finite (closed) with respect to its spatial dimensions*, we should have no need at all of any such boundary conditions. We shall proceed to show that both the general postulate of relativity and the fact of the small stellar velocities are compatible with the hypothesis of a spatially finite universe; though certainly, in order to carry through this idea, we need a generalizing modification of the field equations of gravitation.

OPPOSITE AND FOLLOWING PAGE

String theory—developed largely since Einstein's death— has brought about new theories of how the universe could have begun.

OPPOSITE PAGE

A representation of a recent model of the beginning of the universe from the perspective of string theory and brane theory. As two exhausted branes (multidimensional existences) draw closer to one another they reach across many dimensions to create one or many Big Bangs. The wild and cataclysmic contact throws them apart, but in doing so regenerates the latent energies.

OPPOSITE PAGE

*The Final Brane in string
theory represents the end
and the beginning of the
sequence of an unfolding
universe—the big bang that
comes from the big crunch.*

§3. THE SPATIALLY FINITE UNIVERSE WITH A UNIFORM DISTRIBUTION OF MATTER

According to the general theory of relativity the metrical character (curvature) of the four-dimensional space-time continuum is defined at every point by the matter at that point and the state of that matter. Therefore, on account of the lack of uniformity in the distribution of matter, the metrical structure of this continuum must necessarily be extremely complicated. But if we are concerned with the structure only on a large scale, we may represent matter to ourselves as being uniformly distributed over enormous spaces, so that its density of distribution is a variable function which varies extremely slowly. Thus our procedure will somewhat resemble that of the geodesists who, by means of an ellipsoid, approximate to the shape of the Earth's surface, which on a small scale is extremely complicated.

The most important fact that we draw from experience as to the distribution of matter is that the relative velocities of the stars are very small as compared with the velocity of light. So I think that for the present we may base our reasoning upon the following approximative assumption. There is a system of reference relatively to which matter may be looked upon as being permanently at rest. With respect to this system, therefore, the contravariant energy-tensor $T^{\mu\nu}$ of matter is, by reason of (5), of the simple form

$$
\begin{array}{cccc}
0 & 0 & 0 & 0 \\
0 & 0 & 0 & 0 \\
0 & 0 & 0 & 0 \\
0 & 0 & 0 & \rho
\end{array}
\tag{6}
$$

The scalar ρ of the (mean) density of distribution may be *a priori* a function of the space coordinates. But if we assume the universe to be spatially finite, we are prompted to the hypothesis that r is to be independent of locality. On this hypothesis we base the following considerations. As concerns the gravitational field, it follows from the equation of motion of the material point that a material point

$$
\frac{d^2 x_\nu}{ds^2} + \{\alpha\beta, \nu\} \frac{dx_\alpha}{ds} \frac{dx_\beta}{ds} = 0
$$

when g_{44} is independent of locality. Since, further, we presuppose independence of the time coordinate x_4 for all magnitudes, we may demand for the required solution that, for all $x<$,

$$g_{44} = 1 \tag{7}$$

Further, as always with static problems, we shall have to set

$$g_{14} = g_{24} = g_{34} = 0 \tag{8}$$

It remains now to determine those components of the gravitational potential which define the purely spatial-geometrical relations of our continuum $(g_{11}, g_{12}, \ldots g_{33})$. From our assumption as to the uniformity of distribution of the masses generating the field, it follows that the curvature of the required space must be constant. With this distribution of mass, therefore, the required finite continuum of the x_1, x_2, x_3, with constant x_4, will be a spherical space.

We arrive at such a space, for example, in the following way. We start from a Euclidean space of four dimensions, $\xi_1, \xi_2, \xi_3, \xi_4$, with a linear element $d\sigma$; let, therefore,

$$d\sigma^2 = d\xi_1^2 + d\xi_2^2 + d\xi_3^2 + d\xi_4^2 \tag{9}$$

In this space we consider the hyper-surface

$$R^2 = \xi_1^2 + \xi_2^2 + \xi_3^2 + \xi_4^2, \tag{10}$$

where R denotes a constant. The points of this hyper-surface form a three-dimensional continuum, a spherical space of radius of curvature R.

The four-dimensional Euclidean space with which we started serves only for a convenient definition of our hyper-surface. Only those points of the hyper-surface are of interest to us which have metrical properties in agreement with those of physical space with a uniform distribution of matter. For the description of this three-dimensional continuum we may employ the coordinates ξ_1, ξ_2, ξ_3 (the projection upon the hyperplane $\xi_4 = 0$) since, by reason of (10), ξ_4 can be expressed in terms of ξ_1, ξ_2, ξ_3. Eliminating ξ_4 from (9), we obtain for the linear element of the spherical space the expression

$$\left.\begin{array}{l} d\sigma^2 = \gamma_{\mu\nu} d\xi_\mu d\xi_\nu \\[2mm] \gamma_{\mu\nu} = \delta_{\mu\nu} + \dfrac{\xi_\mu \xi_\nu}{R^2 - \rho^2} \end{array}\right\} \tag{11}$$

where $\delta_{\mu\nu} = 1$, if $\mu = \nu$; $\delta_{\mu\nu} = 0$, if $\mu \neq \nu$, and $\rho^2 = \xi_1^2 + \xi_2^2 + \xi_3^2$. The coordinates chosen are convenient when it is a question of examining the environment of one of the two points $\xi_1 = \xi_2 = \xi_3 = 0$.

Now the linear element of the required four-dimensional space-time universe is also given us. For the potential $g_{\mu\nu}$, both indices of which differ from 4, we have to set

$$g_{\mu\nu} = -\left(\delta_{\mu\nu} + \frac{x_\mu x_\nu}{R^2 - (x_1^2 + x_2^2 + x_3^2)}\right) \tag{12}$$

which equation, in combination with (7) and (8), perfectly defines the behavior of measuring-rods, clocks, and light-rays.

———

§ 4. CONCLUDING REMARKS

The above reflections show the possibility of a theoretical construction of matter out of gravitational field and electromagnetic field alone, without the introduction of hypothetical supplementary terms on the lines of Mie's theory. This possibility appears particularly promising in that it frees us from the necessity of introducing a special constant 8 for the solution of the cosmological problem. On the other hand, there is a peculiar difficulty. For, if we specialize (1) for the spherically symmetrical static case we obtain one equation too few for defining the $g_{\mu\nu}$ and $\phi_{\mu\nu}$, with the result that any *spherically symmetrical distribution* of electricity appears capable of remaining in equilibrium. Thus the problem of the constitution of the elementary quanta cannot yet be solved on the immediate basis of the given field equations.

Stephen Hawking

Stephen Hawking is considered the most brilliant theoretical physicist since Einstein. He has also done much to popularize science. His book, *A Brief History of Time*, sold more than 10 million copies in 40 languages, achieving the kind of success almost unheard of in the history of science writing. His subsequent books, *The Universe in A Nutshell*, and *The Future of Spacetime*, with Kip S. Thorne and others, have also been well-received.

He was born in Oxford, England on January 8, 1942 (300 years after the death of Galileo). He studied physics at University College, Oxford, received his Ph.D. in Cosmology at Cambridge and since 1979, has held the post of Lucasian Professor of Mathematics. The chair was founded in 1663 with money left in the will of the Reverend Henry Lucas, who had been the member of Parliament for the University. It was first held by Isaac Barrow, and then in 1663 by Isaac Newton. It is reserved for those individuals considered the most brilliant thinkers of their time.

Professor Hawking has worked on the basic laws that govern the universe. With Roger Penrose, he showed that Einstein's General Theory of Relativity implied space and time would have a beginning in the Big Bang and an end in black holes. The results indicated it was necessary to unify General Relativity with Quantum Theory, the other great scientific development of the first half of the twentieth century. One consequence of such a unification was that he discovered that black holes should not be completely black but should emit radiation and eventually disappear. Another conjecture is that the universe has no edge or boundary in imaginary time.

Stephen Hawking has twelve honorary degrees, and is the recipient of many awards, medals, and prizes. He is a Fellow of the Royal Society and a Member of the U.S. National Academy of Sciences. He continues to combine family life (he has three children and one grandchild) and his research into theoretical physics together with an extensive program of travel and public lectures.

ENDNOTES

NICOLAUS COPERNICUS

1. This foreword, at first ascribed to Copernicus, is held to have been written by Andrew Osiander, a Lutheran theologian and friend of Copernicus, who saw the *De Revolutionibus* through the press.

2. Ptolemy makes Venus move on an epicycle the ratio of whose radius to the radius of the eccentric circle carrying the epicycle itself is nearly three to four. Hence the apparent magnitude of the planet would be expected to vary with the varying distance of the planet from the Earth, in the ratios stated by Osiander. Moreover, it was found that, whenever the planet happened to be on the epicycle, the mean position of the Sun appeared in line with *EPA*. And so, granted the ratios of epicycle and eccentric, Venus would never appear from the Earth to be at an angular distance of much more than 40° from the center of her epicycle, that is to say, from the mean position of the Sun, as it turned out by observation.

3. The three introductory paragraphs are found in the Thorn centenary and Warsaw editions.

4. The "orbital circle" (*orbis*) is the great circle whereon the planet moves in its sphere (*sphaera*). Copernicus uses the word *orbis*, which designates a circle primarily rather than a sphere because, while the sphere may be necessary for the mechanical explanation of the movement, only the circle is necessary for the mathematical.

GALILEO GALILEI

1. "Natural motion" of the author has here been translated into "free motion"—since this is the term used today to distinguish the "natural" from the "violent" motions of the Renaissance. [*Trans.*]

JOHANNES KEPLER

1. For in the *Commentaries on Mars*, chapter 48, page 232, I have proved that this Arithmetic mean is either the diameter of the circle which is equal in length to the elliptic orbit, or else is very slightly less.

2. Kepler always measures the magnitude of a ratio from the greater term to the smaller, rather than from the antecedent to the consequent, as we do today. For example, as Kepler speaks, 2:3 is the same as 3:2, and 3:4 is greater than 7:8.—C. G. Wallis.

3. That is to say, since Saturn and Jupiter have one revolution with respect to one another every twenty years, they are 81° apart once every twenty years, while the end-positions of this 81° interval traverse the ecliptic in leaps, so to speak, and coincide with the apsides approximately once in eight hundred years.—C. G. Wallis

ALBERT EINSTEIN

1. The preceding memoir by Lorentz was not at this time known to the author.

2. *I.e.* to the first approximation.

3. We shall not here discuss the inexactitude which lurks in the concept of simultaneity of two events at approximately the same place, which can only be removed by an abstraction.

4. "Time" here denotes "time of the stationary" and also "position of hands on the moving clock situated at the place under discussion."

5. A. Einstein, Jahrbuch für Radioakt. und Elektronik, 4, 1907.

6. Of course we cannot replace any arbitrary gravitational field by a state of motion of the system without a gravitational field, any more than, by a transformation of relativity, we can transform all points of a medium in any kind of motion to rest.

7. The dimensions of S_1 and S_2, are regarded as infinitely small in comparison with h.

8. See above.

9. L. F. Jewell (Journ. de Phys., 6, 1897, p. 84) and particularly Ch. Fabry and H. Boisson (Comptes rendus, 148, 1909, pp. 688-690) have actually found such displacements of fine spectral lines toward the red end of the spectrum, of the order of magnitude here calculated, but have ascribed them to an effect of pressure in the absorbing layer.

10. Of course an answer may be satisfactory from the point of view of epistemology, and yet be unsound physically, if it is in conflict with other experiences.

11. Eötvös has proved experimentally that the gravitational field has this property in great accuracy.

12. We assume the possibility of verifying "simultaneity" for events immediately proximate in space, or—to speak more precisely—for immediate proximity or coincidence in space-time, without giving a definition of this fundamental concept.

13. D is the mean density of matter, calculated for a region which is large as compared with the distance between neighboring fixed stars, but small in comparison with the dimensions of the whole stellar system.

14. Akad de Sitter. Van Wetensch. te Amsterdam, November 8, 1916.

ACKNOWLEDGEMENTS

Page 6: Moon*runner* Design.

Page 8: Moon*runner* Design.

Page 12: Portrait of Nicolas Copernicus, Detlev van Ravenswaay/Science Photo Library.

Page 14: Geocentric model of Ptolemy; Kupferstich, 1742/Doppelmaier, 1742.

Page 15: The Copernican system; Nicholaus Copernicus: *De RevolutionIbus Orbium Socelestium Libri VI*, Nürnberg, 1543.

Page 16: NASA Johnson Space Center—Earth Sciences and Image Analysis (NASA-ES&IA).

Page 17: Portrait of Ptolemy from a fifteenth-century manuscript; Biblioteca Marciana, Florence.

Page 18: Discussion between Theologian and Astronomer; Alliaco, *Concordantia Astronomiae cum Theologia*. Erhard Ratdolt, Augsburg, 1490.

Page 19: Etching by Jan Luyken; Moravska Gallery, Brno.

Page 20: Nicolaus Copernicus with heliocentric system; Copper engraving, Pierre Gassendi, *Biography of Copernicus*. Paris, 1654.

Page 21: A Christian Philosopher; George Hartgill, *General Calendars*, London, 1594.

Page 22: Moon*runner* Design.

Page 25: Moon*runner* Design.

Page 26: Moon*runner* Design.

Page 28–29: Proof that the Earth is a sphere; Peter Apian, *Cosmographicus Liber*. Antwerp, 1533.

Page 30: NASA Glenn Research Center (NASA-GRC).

Page 31: NASA Johnson Space Center (NASA-JSC).

Page 34: John Feld, PC Graphics Reports.

Page 36: Atlas bearing Universe by William Cunningham; William Cunningham, *The Cosmographical Glasse*, London, 1559.

Page 37: NASA Goddard Space Flight Center (NASA-GSFC)

Page 38–39: Flemish armillary sphere, 1562; Adler Planetarium, Chicago.

Page 42–43: Pair of Spanish one-handed dividers, 1585; Dudley Barnes Collection, Paris.

Page 46: NASA Glenn Research Center (NASA-GRC).

Page 50: Galileo about age 46; Scala, Florence.

Page 52: Galileo; History of Science Collections, University of Oklahoma Libraries.

Page 53: Galileo on Trial; Bridgeman Art Library, London.

Page 54: Florence by Giorgio Vasari; Scala, Florence.

Page 55: University of Padua; Scala, Florence.

Page 56–57: NASA Jet Propulsion Laboratory (NASA-JPL).

Page 59: Title page for Galileo's book *Dialogue Concerning the Two Chief World Systems*; History of Science Collections, University of Oklahoma Libraries.

Page 61: Moon*runner* Design.

Page 63: Moon*runner* Design.

Page 65: Moon*runner* Design.

Page 67: Moon*runner* Design.

Page 68–69: Telescopes designed and built by Galileo—both preserved in Venice; The Mansell Collection, London.

Page 70: Moon*runner* Design.

Page 74: (Taxi) Getty Images.

Page 76–77: Science Photo Library.

Page 78: Steve Allen/Science Photo Library.

Page 84: NASA Johnson Space Center—Earth Sciences and Image Analysis (NASA-JSC-ES&IA).

Page 88: Galileo presenting his telescope to the Doge of Venice; Institute and Museum of the History of Science, Florence.

Page 90–91: Picture of Moon at first quarter drawn by Galileo; History of Science Collections, University of Oklahoma Libraries.

Page 93: Phases of the Earth's Moon—watercolor by Galileo; Biblioteca Nazionale Centrale, Florence.

Page 97: Moon*runner* Design.

Page 98: Kepler; Sternwarte Kremsmunster, Austria.

Page 100: Danish Astronomer Tycho Brahe; Royal Library, Copenhagen.

Page 101: Weilderstadt, Germany; Landesbildstelle Wurttemberg, Stuttgart.

Page 102: University of Tübingen; New York Public Library, New York.

Page 103: Graz; New York Public Library, New York.

Page 104: Johann Kepler's Diagram of the Geometric Relationships underlying God's Scheme for the planetary orbits; Johann Kepler: *Mysterium Cosmographicum*, 1596.

Page 106: First Wife of Kepler; Museum of the Russian Academy of Sciences, St. Petersburg.

Page 107: Johannes Kepler; Dr. Dow Smith in association with Itec Corp, Seattle/Tokyo.

Page 108: Tycho Brahe and Johannes Kepler; *Atlas Coelestis*. Doppelmaier, Nürnberg, 1742.

Page 110: Johannes Kepler; The Library of Congress, Washington D.C.

Page 111: Great Globe; Pramonstratenserkluster in Strahov, Prague.

Page 112–113: Moon*runner* Design.

Page 115: Tycho Brahe's quadrant; *Astronomiae Instauratae Mechanica*. Levin Hulsi, Nuremberg, 1598.

Page 116–117: Moon*runner* Design.

Page 120: Moon*runner* Design.

Page 125: Models of Ptolemy, Copernicus, and Tycho Brahe; Johannes Zahn, *Specula Physico-Mathematico-Historica Notabilium ac Mirabilium Scientiarum; in qua Mundi Mirabilis Oeconomia*. Nürnberg, 1696.

Page 126: Drawing of Copernican system by Thomas Digges; Thomas Digges, 1576.

Page 128–129: John Feld, PC Graphics Report.

Page 130: World system determined by geometry of the regular solids; Johann Kepler, *Harmonices Mundi Libri*. Linz, 1619.

Page 134: Tycho Brahe's quadrant and mural at Uraniborg; *Astromias Instauratae Mechanica*. Wandesburg, 1598.

Page 136–137: The world system of Tycho Brahe; Tycho Brahe, *De Mundi Aethere: Recentioribus Phaenomicum*. Uraniborg, 1599.

Page 139: NASA Jet Propulsion Laboratory (NASA-JPL).

Page 140–141: Moon*runner* Design.

Page 142: The universe as a monochord by Robert Fludd; *Utriusque Cosmic…Historia* (4 vols.). Oppenheim, 1617–19.

Page 143: Frontpiece, Franchino Gafori, *Practica Musicae*. Milan, 1496.

Page 144: Universe as a harmonious arrangement based on the number 9; Anthansius Kircher, S.J., *Musurgia Universalis* (2 vols.). Rome, 1650.

Page 146: Isaac Newton; Getty Images

Page 149: Frontpiece, Giovanni-Battista Riccioli, S.J, *Almagestum Novum Astronomiam Veterum Novamque Complectens* (2 vols.). Bologna, 1671.

Page 150: Isaac Newton age 12; Lincolnshire County Council.

Page 151: Artist's impression of the apple falling on Newton's head; Mikki Rain, Science Photo Library.

Page 152: Newton fascinated with the properties of the prism; The Mansell Collection, London.

Page 154: Sir Isaac Newton at the age of 77; Zeichung von William Stukeley.

Page 155: Newton, color print 1795 by William Blake; Tate Gallery, London.

Page 156–157: The Principa; Tessa Musgrave, National Trust Photographic Library.

Page 158: Cartoon (18th c.) lampooning Newton's theory of gravity; British Library, London.

Page 161: Sir Issac Newton; National Portrait Gallery, London.

Page 163: Moonrunner Design.

Page 166–168: Diagram of a reflecting telescope by Sir Issac Newton; Prepared by Dr. Dow Smith in association with Itec Corporation, Seattle/Tokyo.

Page 171: Moonrunner Design.

Page 174: Moonrunner Design.

Page 176–177: English telescope, c.1727–1748; Private collection.

Page 179: Moonrunner Design.

Page 182: Moonrunner Design.

Page 186–188: Moonrunner Design.

Page 190: Moonrunner Design.

Page 193: Moonrunner Design.

Page 194: Albert Einstein in 1920; Albert Einstein.'

Page 196: Young Albert Einstein; Einstein Archives, New York.

Page 199: Albert Einstein with his family; Einstein Archives, New York.

Page 200: Albert Einstein in Berlin; Schweizerische Landesbibliothek, Bern.

Page 201: Albert Einstein with Charlie Chaplin at the Premier of City Lights; Ullstein Bilderdienst.

Page 202: Albert Einstein; The Jewish National and University Library, Jerusalem.

Page 204: Moonrunner Design.

Page 207: Moonrunner Design.

Page 212: Moonrunner Design.

Page 214: Moonrunner Design.

Page 219: Moonrunner Design .

Page 221: Moonrunner Design.

Page 223: Moonrunner Design.

Page 224–225: Moonrunner Design.

Page 226–227: Moonrunner Design.

Page 230–231: Moonrunner Design.

Page 233: Moonrunner Design.

Page 235: Moonrunner Design.

Page 236–237: Moonrunner Design.

Page 240–243: Moonrunner Design.

Page 244: Moonrunner Design.

Page 247: Moonrunner Design.

Page 250: Stephen Hawking in 2001; Stewart Cohen.

Jacket: Moonrunner Design.